CISM Courses and Lectures

Series Editors:

The Rectors
Friedrich Pfeiffer - Munich
Franz G. Rammerstorfer - Wien
Elisabeth Guazzelli - Marseille

The Secretary General
Bernhard Schrefler - Padua

Executive Editor
Paolo Serafini - Udine

The series presents lecture notes, monographs, edited works and
proceedings in the field of Mechanics, Engineering, Computer Science
and Applied Mathematics.
Purpose of the series is to make known in the international scientific
and technical community results obtained in some of the activities
organized by CISM, the International Centre for Mechanical Sciences.

International Centre for Mechanical Sciences

Courses and Lectures Vol. 548

For further volumes:
www.springer.com/series/76

Sergio Chibbaro · Jean Pierre Minier

Editors

Stochastic Methods in Fluid Mechanics

 Springer

Editors

Sergio Chibbaro
Université Pierre Et Marie Curie, Paris, France

Jean Pierre Minier
R&D EDF, Chatou, France

ISSN 0254-1971
ISBN 978-3-7091-1621-0   ISBN 978-3-7091-1622-7 (eBook)
DOI 10.1007/ 978-3-7091-1622-7
Springer Wien Heidelberg New York Dordrecht London

All contributions have been typeset by the authors
Printed in Italy

Printed on acid-free paper

Springer is part of Springer Science+Business Media (www.springer.com)

Preface

Jean-Pierre Minier [*] and Sergio Chibbaro [†]

[*] EDF R&D, MFEE, 6 Quai Watier 78400 Chatou, France
[†] Institut Jean Le Rond D'Alembert, University Paris 6, 4, place jussieu 75252
Paris Cedex 05

A student trying to get first ideas on, for example, reactive two-phase flow modelling by browsing through available literature is easily faced with a number of terms such as PDF approaches, Monte Carlo methods, stochastic models, Kinetic or Langevin equations, etc., which, further compounded by the regrettable but usual opposition between Eulerian and Lagrangian points of view, create a rather confused picture. Then, even as acronyms become more and more familiar (when PDF is decoded as Probability Density Function thus pointing to the fact the aim of a PDF method is to simulate the probablility density function of some variables), the relations between these different aspects are not straightforward. By opening reference textbooks, the situation is not immediately improved as names such as Liouville, Fokker-Planck or Master equations seem to appear without necessarily any clear order or hierarchy between them. Furthermore, identical concepts can be referred to quite differently in Physics and in Mathematics: for instance, what is usually (but loosely) named Langevin equations in physical applications corresponds to stochastic differential equations of a stochastic diffusion process in mathematical works. Therefore, finding one's way in this attractive but intricate landscape may not be an uneventful journey...

This situation can be explained by historical developments. Indeed, stochastic processes were first introduced in natural sciences and engineering through the work of Einstein on Brownian motion in 1905. Statistical analysis had started before with Maxwell and Boltzmann but a specific consideration of 'fluctuations' and of stochastic effects can be said to have been initiated by Einstein's work. At about the same period, further works, in particular by Langevin, Smoluchowski and others, extended this first breakthrough to the broader class of general diffusion processes, opening the way to one of the most fruitful branch of modern science with developments in chemistry, condensed-matter, statistical mechanics, applied physics, mechanics and engineering. These first steps were made in the absence of any strict theory and it is only a few years later that mathematical developments started to cast these notions into a rigorous framework, in particular through the work of Wiener (1921). Later on, probability was set on a rigorous axiomatic basis by Kolmogorov in 1933 and, still later, Ito (1942) proposed the first clear definition of a stochastic integral. Like a delayed echo in Physics, another formulation of stochastic integrals was later put forward by Stratonovich (1965) and this illustrates the corresponding but not always well connected developments made from a mathematical or a physical point of view. Since these early attempts, stochastic processes have been used in several areas of science though they are often introduced in different frameworks which have, apparently, little in common. Stochastic processes are a current subject of intense research activity in mathematics, with a typical outlook towards Finance and Economics issues, but the highly involved and mathematically-oriented formulation may render their presentation somewhat difficult to grasp for researchers with a more physical viewpoint. On the other hand, in physics (whether it is theoretical, applied physics or in some domains of Engineering) stochastic processes have been also developed but mostly from each separate problems at hand. For example, they have been applied in chemical studies, under the name of a Master Equation which deals with discrete processes, or in Fluid Turbulence and for Combustion and Reactive Flows, under the term of PDF Methods which handles diffusion processes and are related to Fokker-Planck PDF models. Relations between these developments can be missed and, as a consequence, new ideas are not easy to carry from one field to the next one.

The purpose of the course held at the CISM in Udine (2-6 July 2012) was to provide a general and unified framework in which stochastic processes are presented as ***modelling tools*** for various issues in physics, chemistry and engineering with a particular focus on fluid mechanics. The aim was therefore to develop what can referred to as **Stochastic Modelling** for a whole range of applications. It is a middle-of-the-road approach between rigorous mathematical studies (which aim at proving properties on stochastic processes) and the various physical and engineering studies (which aim at using directly some properties of stochastic processes for specific issues). In that respect, the purpose of the course was to propose a mathematically correct but yet simplified picture of the key aspects and properties of stochastic processes so that they can be manipulated without risks of slipping into mistakes and applied to provide insights into physical situations and practical problems.

In this general framework, a precise objective of the course was to show how stochastic modelling allows us to address various issues related to the derivation of statistical models in a new and powerful way. In particular, one purpose was to show that stochastic modelling introduces naturally a *mesoscopic level* where models can be formulated in a stochastic language which is very helpful for the expression of *macroscopic constitutive relations*. Indeed, in the context of statistical formulations, stochastic modelling provides an interesting frame which, once mathematical aspects have been understood, turn out to be a major help for physical analysis and concerns.

For these reasons, the course started with a presentation of important mathematical issues related to stochastic processes. This is reflected in Chapter 1 of this volume which aims at putting forward key notions of stochastic processes and at clarifying the common points between existing formulations (PDF models, Lagrangian models, Monte Carlo methods, etc). A second specific course was devoted to the development of numerical schemes for practical simulations of stochastic processes. This part is not reproduced within this volume but a detailed account can be found in Peirano et al. (2006). The rest of the course was devoted to illustrating the interest and the range of possible applications in several domains of applied physics and mechanics. Two complete courses concerned the subject of colloidal particle agglomeration and deposition/resuspension. They are not reproduced in this volume but details be found in specific articles (see Mohaupt et al. (2011) for colloidal particle collision and agglomeration) and in a recent comprehensive review article (Henry et al., 2012) on the general issue of particulate fouling. Among the various domains of interest, particular attention was concentrated on the specific issue of polydisperse turbulent two-phase flows which is a subject of great practical interest while proposing fascinating theoretical challenges. Apart from the range of applications that are directly related to it, this subject is a direct illustration of the importance of having physical modelling ideas which are based on a rigorous and well-developed mathematical basis, as emphasised in this Introduction. Indeed, earlier attempts to propose PDF descriptions made interesting and, in that sense, pioneering steps but have failed to set up a rigorous basis and have thus been hampered and even sapped by internal inconsistencies. During the course, detailed presentations were proposed where the same issue is addressed based on a more rigorous mathematical foundation (such as the choice of a Markovian description) and on the key physical notions of scale-separation and of fast/slow variables which are gathered in the central notion of the choice of a proper mesoscopic description and, more specifically, on the importance of *a relevant choice of the state vector*. This is proposed in detail in Chapter 2. Another presentation dealing with the same subject of polydispere two-phase flows was made by Pr. Fox who developed Quadrature-based Moments methods which, as indicated by their names, are trying to capture moments in relation with a stochastic description and are, in that sense, interesting illustrations of how stochastic modelling can be used for a consistent formulation of macroscopic quantities. These recent ideas and developments are detailed in Chapter 3. In illuminating presentations, Pr. Succi described first how stochastic ideas were used in the development of lattice-gas methods and, later, with the more powerful Lattice-Bolztmann Methods (LBM) and, second, how methods such as Dissipative Particle Dynamics (DPD) provide an interesting modelling framework at a mesoscopic level which bridges the gap between molecular and hydrodynamical levels. These subjects are presented in detail in Chapter 4.

Bibliography

E. Peirano, S. Chibbaro, J. Pozorski and J.-P. Minier, *Mean-field/PDF numerical approach for polydispersed turbulent two-phase flows*, Progress in Energy and Combustion Science, Vol. 32, pages 315-371, 2006.

M. Mohaupt, J.-P. Minier and A. Tanière, *A new approach for the detection of particle interactions for large-inertia and colloidal particles in turbulent flows*, International Journal of Multiphase Flows 37 (2011), 746-755.

C. Henry, J.-P. Minier and G. Lefèvre, *Towards a description of particulate fouling: from single phase deposition to clogging*, Advance in Colloid and Interface Science, 185-186 (2012), 34-76.

CONTENTS

Mathematical background on stochastic processes

Jean-Pierre Minier [*] and Sergio Chibbaro [†]

[*] EDF R&D, MFEE, 6 Quai Watier 78400 Chatou, France
[†] Institut Jean Le Rond D'Alembert, University Paris 6, 4, place jussieu 75252 Paris Cedex 05

1 Why study mathematical aspects?

As mentioned in the introduction to this volume, the purpose of the formation was mainly to provide an in-depth presentation of stochastic processes as modelling tools. The chosen standpoint is thus a physical point of view. However, it has also been emphasised that, in order to be able to follow the details of specific models and build bridges between different subjects where new ideas related to stochastic modelling can appear, researchers must have a sound knowledge of the mathematical properties of stochastic processes. This represents a middle-of-the-road approach. Indeed, even though the subject is still relatively young, a vast mathematical literature exists on stochastic processes (Arnold, 1974; Klebaner, 1998; Oksendal, 1995; Karatzas and Shreve, 1991) but these works may not be easily accessible to physically-oriented readers. On the other hand, stochastic processes have been used in separated fields of Applied Physics but not always with a clear presentation or resorting to some 'recipes'. Yet, in recent decades, attempts have been made to come up with improved introductions to stochastic processes in the Physics community. In particular, the books by Van Kampen (1992) and especially by Gardiner (1990) marked a first step in addressing the issue from a general perspective (thus the name handbook appearing in Gardiner's title) but are still subject to some critics from a strict mathematical standpoint. In terms of the level of details provided and of mathematical rigour, an improved presentation (in the present authors' opinion) was later made by Ottinger (1996) who, interestingly enough, proposed a introduction to stochastic modelling out of a specific physical issue related to polymer study. Yet, perhaps as the consequence of this very issue, only diffusion stochastic processes were considered. It may also be noted that other presentations can be found in review articles on stochastic methods for fluid dynamics (Pope, 1985; Minier and Peirano, 2001) where, in particular in the latter one, a unified version of so-called PDF equations is proposed. In this chapter, following this middle-of-the-road approach, a mathematical introduction to the main characteristics of stochastic processes is presented.

Before outlining some basic characteristics of stochastic processes, a few words to illustrate where stochastic processes come into play are in order. From a physical point of view, stochastic processes and stochastic differential equations (SDE) appear naturally in systems where a so-called 'white-noise' is acting. At this stage, it is worth making the distinction between two situations.

On the one hand, if we consider a classical mechanical description where a mechanical object is described by a certain state-vector, say X, whose evolution equation is under the influence of an external random force that can be expressed by

$$\frac{dX(t)}{dt} = f(t, X(t), Y(t)) \tag{1}$$

and if the random functions, $Y(t)$, are 'smooth-enough', then the above equations are meaningful and classical calculus, as well as the theory of Ordinary Differential Equations (ODE), are still applicable. In other terms, such noises are referred to as 'coloured noise' and no new theory is needed.

On the other hand, if we consider the same mechanical system but under the influence of a 'white-noise', for example through an evolution equation of the type

$$\frac{dX(t)}{dt} = f(t, X(t)) + g(t, X(t))\, \zeta_t \tag{2}$$

where ζ_t represents a very-fast noise (constant spectral density and **infinite energy** but with a **zero-correlation timescale**), then a new theory is needed to define and handle these equations!

2 An introduction to Stochastic Processes

2.1 Random variables

Simplified presentation. In most physically-oriented textbooks, random variables are often introduced directly and loosely defined as one variable, say X, which can take different possible values, say x, in a corresponding sample space. This sample space can be discrete or continuous, for example $\mathbb{R}$ or $\mathbb{R}^d$ for a d-dimensional vector variable. The distribution of all possible values, x, of the random variable X is measured by introducing the probablity density function (pdf) $p_X(x)$ which is defined by

$$\mathbb{P}\left[\, x \leq X \leq x + dx \,\right] = p_X(x)\, dx. \tag{3}$$

As can be seen, the pdf is a density and only integrated values over a small intervals are meaningful, although $p_X(x)$ is loosely referred to as the probablity to have $X = x$ (which is mathematically meaningless or would always be 0 since the 'length' of the interval is zero in that case!). The pdf is the key quantity to characterise the random variable (in a weak sense) since it gives access to all statistics derived from X, and we have for any chosen function, say Q, that

$$\langle Q(X) \rangle = \int Q(x) p_X(x)\, dx. \tag{4}$$

The approach is extended to joint random variables, say $\mathbf{Z} = (X, Y)$, by introducing their joint probablity density function, $p_{X,Y}(x, y)$, defined as

$$\mathbb{P}\left[\, x_1 \leq X < x_2\, ;\, y_1 \leq Y < y_2 \,\right] = \int_{x_1}^{x_2} \int_{y_1}^{y_2} p_{(X,Y)}(x, y)\, dx\, dy. \tag{5}$$

Once joint random variables are characterised (by their joint pdf), information is retrieved on each separate random variables as marginals

$$p_X(x) = \int p_{(X,Y)}(x, y)\, dy \qquad p_Y(y) = \int p_{(X,Y)}(x, y)\, dx. \tag{6}$$

Then, conditional random variables can also be introduced (from two joint variables) by defining for example the probability distribution of one variable, say Y, but conditioned on the event $X = x$ (accepting the loose notation already mentioned above) as

$$p_{Y|(X=x)} = \frac{p_{(X,Y)}(x, y)}{p_X(x)}. \tag{7}$$

It must be noted that mean conditional statistics of $Y|(X = x)$ are also random variables (basically as a function of the conditioned event) and that we have

$$\langle Q(X, Y)|(X = x) \rangle = \int Q(x, y) p_{Y|X}(y)\, dy \tag{8}$$

from which the unconditional statistics are obtained by further integration over all possible values of the condition

$$\langle Q(X, Y) \rangle = \int \langle Q(X, Y)|(X = x) \rangle\, p_X(x)\, dx. \tag{9}$$

Many related notions can be found in more details in most classical tetxt-books on the subject. Yet, some useful relations (which often serve as starting points for the derivation of PDF equations in sample-space, see Pope (1985); Minier and Peirano (2001)) can be already obtained, for example:

$$\langle\, \delta(X - x)\,\rangle = p_X(x). \tag{10}$$

In spite of its apparently obvious nature, this is an intricate result since the left-hand side of the equation involves the expectation operator, the Dirac distribution (since it is rigourously a distribution rather than a function), the random variable and one of its possible value in its corresponding sample space. A further result involves two-joint random variables and is given by the following relation

$$\langle\, Y\delta(X - u)\,\rangle = \langle\, Y|(X = u)\,\rangle\, p_X(u). \tag{11}$$

However, this simplified and short-cut presentation of random variables can be faced with severe problems when more involved notions come into play. For example, the notion of the different modes of convergence of a series of random variables cannot be understood. Furthermore, conditional random variables can only be understood when we start with two (joint) random variables whereas the real notion can involve only a coarse-grained version of *a single random variable*. This important notion is completely missed by such a simplified presentation. Finally, and this is a key point for our present concern, this simplified presentation is a severe limitation when we want to introduce stochastic processes and, in particular, the natural correspondance between the trajectory and the pdf points of view.

Rigorous presentation. From a rigorous point of view, a random variable X is defined as a measurable function

$$(\Omega, \mathcal{F}, P) \longrightarrow (A, \mathcal{G})$$
$$\omega \longrightarrow X(\omega) \tag{12}$$

and $P_X(B) = P(X^{-1}(B))$ is the probablity measure induced where $B \in \mathcal{G}$ is a subset (or an 'event') of the σ-algebra A.

When for example the arrival ensemble A is equal to $\mathbb{R}$ (or $\mathbb{R}^d$) then, by considering the Borel σ-algebra (generated by intervals) and by assuming a density with respect to the Borel measure dx, we retrieve the simplified notion of probablity density functions

$$\mathbb{P}\left([a, b[\right) = \int_a^b p_x(u)\, du. \tag{13}$$

A very detailed and easily-accessible presentation can be found in the book by Ottinger (1996). It is seen that, from a mathematical point of view, a random variable is defined as the map from an underlying (or reference) probability space, Ω, equipped with a set of subsets that form a σ-algebra A and with a reference probability measure P which is a measure of this reference 'chance' or 'randomness'. Even though this reference probability space is not explicitely detailed, the idea of having a random variable as a measurable function is actually quite natural. As beautifully explained in particular in the first part of Ottinger (1996), the introduction of σ-algebras corresponds to the intuitive notion of the 'level of information' that is resolved by one description (that is by the specific choice of a random variable) and, therefore, corresponds directly to the notion of coarse-grained and fine-grained descriptions of the same object (represented directly here by different σ-algebras, one being contained in the other one).

Having rigorously defined random variables, we are now in a position to introduce and understand the various modes with which a series of random variables can be said to converge to a limit random variable. For that purpose, let us consider $(X_n)_{n\in\mathbb{N}}$ and X, random variables with values in $\Re$ and defined on the same probablity space. Four modes of convergence can be defined:

1. X is the *almost sure* limit of the sequence (X_n), if

$$\mathbb{P}\left(\{\omega||X_n(\omega) - X(\omega)| \to 0 \text{ as } n \to \infty\}\right) = 1$$

2. X is the *mean-square* limit of the sequence (X_n), if

$$\langle |X_n - X|^2 \rangle \to 0 \text{ as } n \to \infty$$

3. X is the *stochastic* limit of the sequence (X_n), if

$$\mathbb{P}\left(\{\omega||X_n(\omega) - X(\omega)| \geq \epsilon\}\right) \to 0 \text{ as } n \to \infty$$

4. X is the limit *in distribution* of the sequence (X_n) if

$$\forall g, \quad \lim_{n\to\infty} \langle g(X_n) \rangle = \langle g(X) \rangle$$

For the first three modes of convergence, it is clear that $(X_n)_{n\in\mathbb{N}}$ and X must be defined on the same probability space (a notion that escapes the simplified presentation) while this is not necessary for the convergence in law or in distribution (fourth mode of convergence). These modes of convergence have been considered in detail, for example in (Ottinger, 1996), and they are essential as they enter important results or concepts in stochastics. The first mode corresponds to the strongest possible mode and is often referred to as convergence 'for each ω' or 'almost everywhere'. The second mode is important since the second-order moment corresponds to the notion of energy in physics and because this mean-square limit plays a central role in the definition of the stochastic integral, as we will see later. Loosely speaking (and leaving aside the third mode), the almost-sure and mean-square modes of convergence can be described as *strong modes of convergence*. The fourth mode of convergence introduces a different notion: in this approach, we are basically interested not directly in the random variables themselves but in the statistics extracted from the random variables. This corresponds to a convergence in the weak sense, or in the distribution sense. The convergence in law is often the relevant mode of convergence in many applications in Physics where approximations of some statistics are being sought. The convergence in law is referred to as a *weak mode of convergence*.

2.2   Stochastic processes

Having briefly recalled some notions about random variables, we now turn to stochastic processes. Following what has been done for random variables, we will first consider a simplified presentation, then put forward some definitions and useful notions about stationary processes, before introducing a more rigorous presentation that is the key to understanding the most important characteristics of stochastic processes.

Simplified presentation. First of all, what is a stochastic process? The answer to that question is actually rather straightforward: a stochastic process is a sequence (or family) of random variables indexed with a continuous parameter which is usually time and is thus written as $(X_t)_{t \in T}$ (where T stands for a time interval).

The second immediate question is: what is the information needed to characterise a stochastic process? This is not such an obvious question and it will be seen that this question is directly related to the important separation between Markovian and non-Markovian processes. Let us now consider the amount of information that is implied by the notion that a given stochastic process is known:

1. At each given time, say t, $X(t)$ is a random variable and (in this simplified presentation) we must know its pdf, that is $p(t; x)$ where x represents the value of the random variable in its sample space. And, of course, this must be valid for each time which means that we must know the family of functions: $p(t; x)$, $\forall t$.

2. For each couple of times, say t_1 and t_2, we must the law of the joint two random variables X_{t_1} and X_{t_2}, which means that we must know the joint pdf $p(t_1, x_1 \,;\, t_2, x_2)$, $\forall t_1$ and t_2.

3. The same if true for any number of chosen times $(t_1, t_2, \ldots, t_N) \ldots$

This reasoning reveals that knowing a stochastic process amounts to the knowledge of the *joint law* of the *joint random variables* $(X_{t_1}, X_{t_2}, \ldots, X_{t_N})$, that is to the knowledge of the *joint law* $p(t_1, x_1; t_2, x_2; \ldots, t_N, x_N)$ for any chosen set of N times and for any values of the times $(t_1, t_2, \ldots, t_N)$! If we regard the joint pdfs as pieces of information, it appear therefore that the amount of information that is needed to characterise a general stochastic process is huge!

As a first consequence, it can be understood that, for a general stochastic process, we can always define or construct the one-time pdfs, $p(t, x)$, $t \in T$ but, without any specific property, this information is not sufficient to characterise the stochastic process.

A note on stationary processes. Before going into a more rigorous presentation of stochastic processes, it is worth considering the subclass of stationary processes since they allow to introduce key physical notions such as the integral timescale and the energy spectrum.

By definition, a stochastic process is said to be stationary if its law is invariant by a shift of time. Since we have just seen that the law of a stochastic process is the joint pdf of the joint random variables $(X_{t_1}, X_{t_2}, \ldots, X_{t_N})$ for any value of N and for any choice of the times $(t_1, t_2, \ldots, t_N)$, this condition can be formulated as: $\forall T, \forall N$ and $\forall (t_1, t_2, \ldots, t_N)$

$$p(t_1 + T, x_1; t_2 + T, x_2; \ldots; t_N + T, x_N) = p(t_1, x_1; t_2, x_2; \ldots; t_N, x_N). \quad (14)$$

As a consequence, the one-time statistics of a stationary process, such as the mean value $\langle X(t) \rangle$ or any moment $\langle X^n(t) \rangle$, are constant while correlations between different times depend only on time lags. One key example is the auto-correlation of the process $\langle X(t) X(t+s) \rangle$ which is only a function of the time lag

$$R(s) = \langle X(t) X(t+s) \rangle.$$

Since one-time statistics are constant, the auto-correlation coefficient is also a function only of the time lag

$$\rho(s) = \frac{\langle X(t) X(t+s) \rangle}{\sqrt{\langle X^2(t) \rangle \langle X^2(t+s) \rangle}} = \frac{\langle X(t) X(t+s) \rangle}{\langle X^2 \rangle}.$$

The integral timescale of the process is then defined as the integral of the auto-correlation coefficient over all (positive) time lag

$$T = \int_0^{+\infty} \rho(s) \, ds. \quad (15)$$

A typical shape of the auto-correlation coefficient is dispayed in Fig. 1, where the dotted square indicates the rectangle having the same size as the integral below the curve of the auto-correlation coefficient. The parameter λ represents the time where the osculating parabola intersects the time coordinate and is related to the value of the Taylor timescale, along with several properties which can be found in more detailed presentations (Pope, 1985, 2000; Minier and Peirano, 2001).

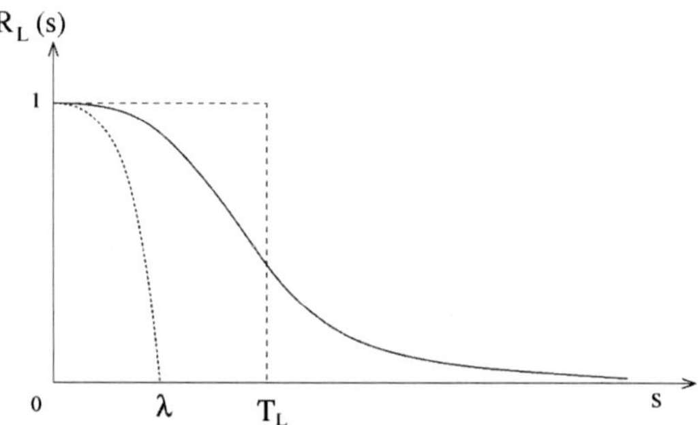

Figure 1. Typical shape of the auto-correlation function of a stationary process, with T_L indicating the integral timescale

The integral timescale is an essential characteristic since it measures the 'memory' of the stochastic process and allows us to give a precise meaning to the intuitive notion of small and large time lags. Indeed, when $s \ll T$, then two successive values of the process $X(t)$ and $X(t+s)$ are highly correlated (since $\rho(s) \simeq 1$) whereas when $s \gg T$, then $X(t)$ and $X(t+s)$ are nearly uncorrelated (since $\rho(s) \simeq 0$). For a stationary process, the integral timescale T appears therefore as the reference time during which successive values remain correlated. Then, with respect to a fixed time T_0, a stationary process having an integral timescale T much smaller than T_0 will be seen as a rapidly-changing process, or *a fast process*, while a stationary process with an integral timescale T much larger than T_0 will be regarded as *a slow process*.

Another important notion is the energy spectrum of a stationary process which is defined as the Fourier-transform of the auto-correlation function

$$E(\omega) = \frac{1}{\pi} \int_{-\infty}^{+\infty} R(s) e^{-i\omega s}\, ds = \frac{2}{\pi} \int_{0}^{+\infty} R(s) \cos(\omega s)\, ds. \qquad (16)$$

By writing the inverse Fourier-transform we have that

$$R(s) = \int_0^{+\infty} E(\omega)e^{i\,\omega s}\,d\omega \tag{17}$$

and, in particular, the value of the origin yields the important relation

$$R(0) = \langle X^2(t)\rangle = \int_0^{+\infty} E(\omega)\,d\omega \tag{18}$$

which shows that $E(\omega)$ represents the density of the energy ($\langle X^2(t)\rangle$) in its wave-function space.

Rigorous presentation. One of the interests of having introduced a rigorous definition of random variables is that it paves the way for a straightforward definition of stochastic processes. As already written above, a stochastic process is a family of random variables indexed by a parameter which is usually time and, thus, appears as a function of two variables

$$T \times (\Omega, \mathcal{F}, P) \longrightarrow (A, \mathcal{G})$$
$$(t, \omega) \longrightarrow X(t, \omega) \tag{19}$$

Based on this rigorous definition, it is now clear that there are two possible ways to handle a stochastic process by fixing one variable and studying the variations with respect to the second variable. Thus, for the *same stochastic process* X_t we can make the following choices:

1. we can first fix the variable t indicating time. Then, for each $t \in T$, we have now a function

$$\omega \longrightarrow X_t(\omega) = X(t, \omega)$$

which shows that $X_t(.)$ is a random variable with a pdf $p(t, x)$. From this first point of view, we are thus led to handle the family of the pdfs, as time evolve, and search for the PDF equation satisfied by $p(t, x)$ in the corresponding sample space;

2. we can also decide to fix the other variable ω which indicates a subset of the sigma algebra of the probablity space on which the stochastic process is defined. This apparently abstract notion represents in fact one 'elementary event', or one outcome, of the underlying randomness governing the process. Then, for each fixed $\omega \in \Omega$, we have now a function of time only

$$t \longrightarrow X_\omega(t) = X(t, \omega)$$

Each of these functions define what is called a *trajectory*, or a *sample path*, of the stochastic process. From this second point of view, we are led to handle a number of trajectories (corresponding to various choices or outcomes of the elementary events ω). This second point of view is the one followed in Monte Carlo methods and, in particular, in the characterisation of a stochastic process by dynamical Monte Carlo approaches.

These two different points of view define what can referred to as 'the PDF point of view' and the 'trajectory point of view' (Minier and Peirano, 2001) and, from the rigorous definition of a stochastic process, it is now obvious that they correspond to two descriptions of the *same mathematical object*, namely the stochastic process itself, and not characterisations of two different processes.

2.3   Markov processes

We have seen that the information required to characterise a general stochastic process is huge. As a consequence, not much can be said about general stochastic processes and attention has been focused on classes of stochastic processes for which the information can be reduced to a tractable level. In that sense, the most important class of stochastic process is made up by Markov processes.

A markov process corresponds to a process where 'the present is enough to predict the future'. In terms of conditional pdfs, this statement is translated into the definition of a Markov process

Definition 2.1. A stochastic process X_t is a Markov process if for all n and for any chosen set of times $(t_1, t_2, \ldots, t_n)$, we have that

$$p(t_n, x_n \,|\, (t_1, x_1; t_2, x_2; \ldots; t_{n-1}, x_{n-1})) = p(t_n, x_n \,|\, t_{n-1}, x_{n-1}). \tag{20}$$

In the above formula, it is seen that (t_{n-1}, x_{n-1}) represents the present state while (t_n, x_n) represents the future and $(t_1, x_1; t_2, x_2; \ldots, t_{n-2}, x_{n-2})$ the past. The Markov property is often described by stating that the future is independent from the past, but a better formulation is to say that the past is actually contained in the present.

Though surprising at first, the Markov property is actually a rather natural concept and a direct extension of classical mechanics. Indeed, in classical mechanics (say Lagrange Mechanics), one expresses the evolution in time of a given mechanical object by giving an initial condition and its time rate of change. This is translated in an Ordinary Differential Equation (ODE) by saying that, if we know the initial condition and the time rate of change, then the whole history of a mechanical object is known or, in other terms, that the past is contained in the present.

The fundamental property of Markov processes is that the huge amount of information needed to characterise a stochastic process is reduced to the knowledge of just two functions: an initial distribution at a given initial time t_0, $p(t_0, x_0)$, and the transitional density $p(t, x \,|\, s, y)$ (for $t \geq s$). Following what has just been said about classical mechanics, it is seen that the initial distribution represents the extension of the initial condition in ODE while the transitional pdf represents the equivalent of the time rate of change.

It is worth showing that the knowledge of the two functions, $p(t_0, x_0)$ and $p(t, x \,|\, s, y)$, added to the Markov property is enough to generate all information about the stochastic process. Indeed, we can first obtain all one-time pdfs at times $t \geq t_0$ since

$$p(t, x) = \int p(t, x \,|\, t_0, x_0)\, p(t_0, x_0)\, dx_0. \tag{21}$$

Then, the transitional pdf gives access to any two-time pdfs

$$p(t_2, x_2; t_1, x_1) = p(t_2, x_2 \,|\, t_1, x_1)\, p(t_1, x_1). \tag{22}$$

So far, these relations are valid for any stochastic process and do not rely on the Markov property which only comes into play when we consider three-time pdfs and, more generally, n-time pdfs

$$\begin{aligned}
p(t_3, x_3; t_2, x_2; t_1, x_1) &= p(t_3, x_3 \,|\, (t_2, x_2; t_1, x_1))\, p(t_2, x_2; t_1, x_1) \\
&\quad \text{(Markov property)} \\
&= p(t_3, x_3 \,|\, t_2, x_2)\, p(t_2, x_2; t_1, x_1) \\
&= p(t_3, x_3 \,|\, t_2, x_2)\, p(t_2, x_2 \,|\, t_1, x_1)\, p(t_1, x_1).
\end{aligned} \tag{23}$$

This last relation is easily extended to the general case of n-time pdfs, showing that at, any discrete time, the joint pdf is obtained through a chain rule using only the initial pdf and the transitional pdf whose name is justified since it *carries the information* from one time to another:

$$\begin{aligned}
p(t_n, x_n; t_{n-1}, x_{n-1}; \ldots; t_1, x_1) &= p(t_n, x_n \,|\, t_{n-1}, x_{n-1}) \\
&\quad \times p(t_{n-1}, x_{n-1} \,|\, t_{n-2}, x_{n-2}) \times \ldots \times p(t_2, x_2 \,|\, t_1, x_1)\, p(t_1, x_1).
\end{aligned} \tag{24}$$

This is a fundamental property which shows that, apart from an initial condition, the huge amount of information needed to characterise a stochastic process is fully retrieved by one key function: the transitional pdf. Thus, for a Markov process, knowing the process is reduced to knowing a single pdf function which can be obtained by a PDF equation or by the simulation of a large number of trajectories. However, it must be emphasised that this is only true for Markov processes. For a non-Markov process, one can always consider the transitional pdf, $p(t, x \,|\, s, y)$ (for $t \geq s$) but, in that case, this function is not sufficient to characterise the stochastic process: it is then an incomplete description of the process itself! In practice, this means that, to describe a given object, the art of modelling starts by trying to choose relevant state-vectors which can be regarded as Markov processes with a good approximation rather than developing advanced and complex closure methods for a given and fixed state-vector which has very little chance of being a Markov process. It will be seen that this question and the whole issue of choosing a relevant state-vector is at play in two-phase flow modelling.

For Markov processes, the fundamental function is thus the transition pdf which can be shown to satisfy the non-linear equation (Gardiner, 1990; Ottinger, 1996), known as the Chapman-Kolmogorov equation:

$$p(t, x \,|\, t_0, x_0) = \int p(t, x \,|\, t_1, x_1)\, p(t_1, x_1 | t_0, x_0)\, dx_1. \tag{25}$$

More elaborate and important relations are obtained by considering the so-called Infinitesimal Operator which is defined by

$$\mathcal{L}_t g(x) = \lim_{dt \to 0} \frac{\langle (g(X_{t+dt})|X_t = x) \rangle - g(x)}{dt} \tag{26}$$

where it is seen that the Infinitesimal Operator 'mesures' the effect of a conditional increment of the (Markov) stochastic process over a test function g (thus characterising these increments in a weak sense). As indicated by its name, this operator $\mathcal{L}_t$ generates the properties of the stochastic process. In particular, it can be shown (Arnold, 1974; Gardiner, 1990; Ottinger, 1996) that the transition pdf $p(t, x \,|\, s, y)$, which is a function of two sets of variables the 'forward ones' (t, x) and the 'backward ones' (s, y), is the solution of two fundamental equations. When we consider (t, x) as being fixed and the transition pdf as a function of (s, y), then $p(t, x \,|\, s, y)$ is the solution of the following equation

$$\begin{cases} \dfrac{\partial p}{\partial s} + \mathcal{L}_s p = 0 \\[2mm] \text{end cd. } \ p(t, x \,|\, s, y) = \delta(x - y) \quad s \to t \end{cases} \tag{27}$$

which is known as the *Kolmogorov backward equation*. When we consider (s, y) as being fixed and the transition pdf as a function of (t, x), then

$p(t, x \,|\, s, y)$ is the solution of the following equation

$$\begin{cases} \dfrac{\partial p}{\partial t} = \mathcal{L}_t^* p \\ \text{initial cd. } p(t, x \,|\, s, y) = \delta(x - y) \quad t \to s \end{cases} \tag{28}$$

which is known as the *Kolmogorov forward equation* and where $\mathcal{L}_t^*$ stands for the adjoint operator of $\mathcal{L}_t$. From a physical point of view, we are often more interested in the second equation since the problem we are faced with can usually be formulated as: given an initial state and a certain evolution, where do we have a chance to end up? However, the first formulation finds also a natural place in many applications (finance, inverse problems, ...). One noteworthy aspect is that the two equations are obtained, as adjoints, from the *same operator* and, therefore, any modelling through the choice of a particular operator $\mathcal{L}_t$ yields naturally information on *both* the forward and backward problems.

2.4   Two key processes: the Poisson and the Wiener processes

We have seen that, starting from general stochastic processes, attention is focused on the important class of Markov processes. Then, within this class, it is worth making a distinction between two branches or subclasses: processes whose trajectories remain constant but jump at random instants and processes with continuous trajectories and infinitesimal changes within each time step. The first branch corresponds to processes whose trajectories have a small probability to change in an infinitesimal time interval but then evolve with a finite-size jump if a change happens. The second branch corresponds to processes whose trajectories are continuous, therefore with small changes over an infinitesimal time interval, but these changes occur with probability one in any time interval. For each of these two main categories, one process stands out as the canonical example over which generalisations can be built: the Poisson process for jump processes and the Wiener process for diffusion processes.

In the applications with which we are mostly concerned in the present course (such as transport, diffusion, dispersion), we are naturally inclined to consider and handle stochastic processes belonging to the second branch mentioned above, and this will be developed below with a detailed presentation of so-called stochastic diffusion process. It will be seen as the end of this chapter that these two categories are not necessarily separated and can be gathered by considering jump-diffusion processes. However, a few words on the Poisson process itself are first in order.

The Poisson process. The Poisson process $N(t)$ is the basic process that counts random discrete events with no simultaneous multiple occurences being possible (there is a small but non-zero probability to have one jump but multiple jumps at one time are not possible). The trajectories of the Poisson process are therefore piecewise constant with jumps (having a step of one unity in the standard Poisson process) occuring at random times. Some key properties are illustrated in Fig. 2.

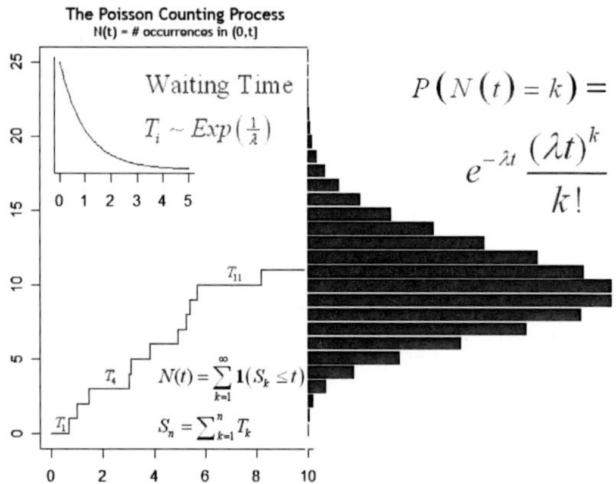

Figure 2. Some properties of the Poisson process $N(t)$: example of one trajectory of the process jumping at random times T_i which follow an exponential distribution and the resulting pdf of $N(t)$ at each time t being a Poisson distribution.

The Poisson process has a number of key properties which are worth mentioning (Gardiner, 1990; Klebaner, 1998):

- the increments, $\Delta N(t) = N(t+\Delta t) - N(t)$, of the Poisson process are stationary and independent;
- a Poisson process is characterised by its intensity, λ, which is the mean value of the number of events occuring per unit time: the number of events in a time interval $[t, t + \Delta t]$ is a Poisson random variable

$$\mathbb{P}[\,\Delta N(t) = k\,] = \frac{(\lambda \Delta t)^k}{k!} e^{-\lambda \Delta t} \tag{29}$$

from which it derives that the mean and variance of the increments are identical and linear with respect to the time increment Δt

$$\langle \Delta N(t) \rangle = \lambda \Delta t \qquad \langle (\Delta N(t) - \langle \Delta N(t) \rangle)^2 \rangle = \lambda \Delta t. \tag{30}$$

- on any finite interval, the individual events are uniformly distributed;
- The waiting times (the time intervals between successive random events or occurences) are random variables which follow an exponential distribution with parameter λ

$$\mathbb{P}[\,T_i = t\,] = \lambda e^{-\lambda t} \quad \Rightarrow \langle T_i \rangle = \frac{1}{\lambda} \tag{31}$$

- the reference timescale is thus $1/\lambda$ and by taking a time interval Δt much smaller than this reference timescale ($\lambda \Delta t \ll 1$), we retrieve the usual presentation of the statistics of the increments

$$\begin{aligned}
\mathbb{P}[\,\Delta N(t) = 0\,] &\simeq 1 - \lambda \Delta t \\
\mathbb{P}[\,\Delta N(t) = 1\,] &\simeq \lambda \Delta t \\
\mathbb{P}[\,\Delta N(t) = k\,] &= 0 \ (k \geq 2)
\end{aligned} \tag{32}$$

The standard Poisson process is the building block for generalised Poisson processes where the jumps can take random discrete values (Gardiner, 1990; Klebaner, 1998). As such, it represents the main process to simulate discrete jump events. Given the fundamental discrete nature of elementary Physical processes (Nature is made of atoms; the discrete jumps involved in Quantum Mechanics; etc.), this has led to consider the corresponding PDF equation as a *Master Equation*, from which diffusion processes would only appear as coarse-grained approximations (Van Kampen, 1992). However, in terms of stochastic modelling, this appear as misleading since it is based on the point of view of one application (even if continuum fluid mechanics is indeed always a coarse-grained version of discrete molecular events ... but can also be regarded as a fundamental starting point for further derivations), leading to possible confusion concerning the status and the relations between various PDF equations. Recent attempts have been made to clarify this issue (Minier and Peirano, 2001) and it is emphasised here that, in fact, jump processes and diffusion processes are stochastic models for two different properties of the phenomena which are modelled (without considering whose phenomenon is more fundamental than the other ones). They can be chosen separately or handled together, as will be recalled at the end of this chapter, defining a stochastic description that encompasses the two key processes (Gardiner, 1990; Minier and Peirano, 2001).

The Wiener process. In Physics, the Wiener process represents Brownian motion (first introduced in an heuristic way by Einstein in 1905 and later put on solid mathematical grounds by Wiener in 1921) while it is the cornerstone of the construction of stochastic differential equations (Arnold, 1974;

Oksendal, 1995). Brownian motion has been the subject of comprehensive studies (Karatzas and Shreve, 1991) detailing many specific properties. In the present context, we will define the Wiener process by the following properties:

  (i) the process has independent increments: $(W_{t_3} - W_{t_2})$ and $(W_{t_1} - W_{t_0})$ are independent when $t_0 < t_1 < t_2 < t_3$

  (ii) the trajectories of the process are continuous functions (almost everywhere)

  (iii) the increments of the Wiener process $(W_{t_2} - W_{t_1})$ are Gaussian random variables, centered and with a variance equal to $(t_2 - t_1)$

The Wiener process has fundamental properties:

- it is a Gaussian, Markov process, with mean and covariance

$$M(t) = \langle W_t \rangle = 0 \quad C(t, t') = \langle W_t W_{t'} \rangle = \min(t, t') \tag{33}$$

- the transitional density is a Gaussian pdf

$$p(t, x \,|\, t_0, x_0) = \frac{1}{\sqrt{2\pi(t - t_0)}} \exp\left(-\frac{(x - x_0)^2}{2(t - t_0)} \right) \tag{34}$$

- the transitional density $p(t, x \,|\, t_0, x_0)$ is the solution of the heat equation, revealing the diffusive nature of the pdf evolution

$$\frac{\partial p}{\partial t} = \frac{1}{2} \frac{\partial^2 p(x)}{\partial x^2} \quad \text{with} \quad p(t, x \,|\, t_0, x_0) = \delta(x - x_0) \quad t \to t_0 \tag{35}$$

- the increments dW_t are stationary and independent

$$\langle dW_t \rangle = 0, \quad \langle (dW_t)^2 \rangle = dt, \quad \langle (dW_t)^n \rangle = o(dt) \tag{36}$$

- the trajectories are continuous but non-differentiable at any point! Furthermore, trajectories have unbounded total variations in any interval!

The last three properties reveal the equivalence between the smooth evolution in sample space of the pdf (through a pure diffusion equation) and the wildly fluctuating behaviour (at any scale!) of the trajectories. The fact that the variance of the increments $\langle (dW_t)^2 \rangle$ is linear in dt instead of scaling with dt^2 is an indication that the trajectories are nowhere differentiable or, in other terms, that the 'velocity' of a Brownian particle is infinite at any instant. From a physical point of view, it is precisely the infinitive 'velocity', or the 'infinitively fast' behaviour of each trajectory (or sample or 'molecule'), that leads to a smooth but irreversible diffusive evolution of the pdf and, therefore, of any moments at a continuum level, see Minier and Peirano (2001, chap. 4). The non-differentiability and infinite total variation of the trajectories of the Wiener process have deep consequences for the construction of the stochastic integral, as will be seen below. A typical example of a trajectory of the Wiener process is displayed in Fig. 3

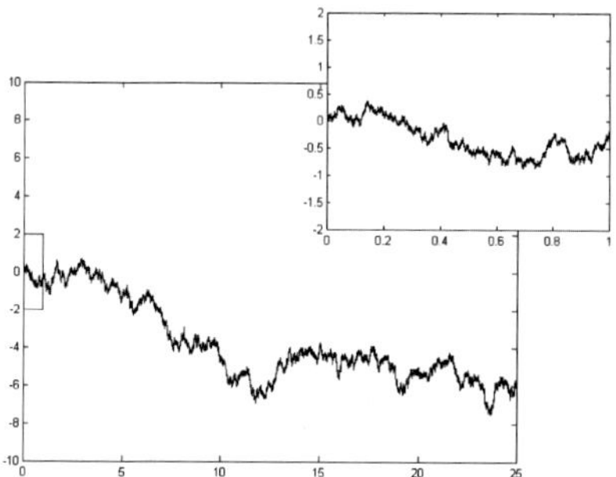

Figure 3. One trajectory of the Wiener process showing a continuous but ragged shape. The window, where a zoom of the trajectory is displayed, indicates the self-similar nature and the (infinitively) fast fluctuations of the trajectories of the Wiener process.

2.5   Stochastic Differential Equations

Motivation.   Before moving into the mathematical aspects of the definition of stochastic integrals, it is worth outlining the motivation and the main ideas from a more physical point of view. For that purpose, following what has been put forward in the introduction of this chapter, we recall that the aim is to give sense to equations where a white-noise ζ_t is acting

$$\frac{dX_t}{dt} = A(t, X_t) + B(t, X_t)\,\zeta_t. \tag{37}$$

We can consider the historical example of the motion of Brownian particles in Langevin's description (see Gardiner (1990, chap. 1))

$$\frac{dV_t}{dt} = -\alpha\,V_t + F_t \tag{38}$$

where the total force acting on a Brownian particle (from molecular collisions) is written as the sum of a friction term and a random term. From an initial condition, $V_0 = 0$, an heuristic integration gives

$$V_t = \int_0^t e^{-\alpha(t-t')}\,F(t')\,dt' = e^{-\alpha t}\int_0^t e^{\alpha t'}\,F(t')\,dt' \tag{39}$$

and from the chaotic nature of molecular impacts (at the scale of a Brownian particle which is assumed to be much larger than the fluid molecules and having a typical timescale much larger than intercollision times) we expect that $\langle F(t')\rangle = 0$ which implies that $\langle V_t\rangle = 0$. From thermodynamic considerations, we also expect that the energy $\langle V_t^2\rangle$ of the Brownian particles reaches an equilibrium value (since the particles are in contact with a heat bath). The particle energy is given by

$$\langle V_t^2\rangle = e^{-2\alpha t}\int_0^t dt' \int_0^t e^{-\alpha(t'+t")}\,\langle F(t')F(t")\rangle\, dt". \tag{40}$$

The first idea is to consider F_t as a stationary, Gaussian, process with independent values at each time but with a finite value:

$$\langle F(t)F(t')\rangle = 0 \quad \text{if } t \neq t' \tag{41}$$
$$= 1 \quad \text{if } t = t' \tag{42}$$

This irregular process can be approximated by a series of 'regular' stationary Gaussian processes Y_t^n

$$\langle Y_t^n\rangle = 0 \tag{43}$$
$$\langle Y_t^n Y_{t'}^n\rangle = e^{-n|t-t'|} \quad (n \to \infty) \tag{44}$$

from which the particle energy is then approximated by

$$\langle (V_t^n)^2\rangle = e^{-2\alpha t}\int_0^t dt' \int_0^t e^{-\alpha(t'+t")}\, e^{-n|t'-t"|}\, dt". \tag{45}$$

However, we end up with the result that $\langle (V_t^n)^2\rangle \to 0$ when $n \to \infty$. Since V_t is a Gaussian process, with a mean $\langle V_t\rangle = 0$ and a variance $\langle V_t^2\rangle = 0$, this implies that the approximation of the noise yields that $V_t = 0$! This is of course a ludicrous result from a physical point of view, showing that the assumption that the noise $F(t)$ could be thought of as a process with finite energy and no memory is not acceptable. We are thus led to consider that the energy of the random term is infinite, in such a way that

$$\langle F(t)F(t')\rangle = \alpha_B\,\delta(t - t') \tag{46}$$

Using this form of the auto-correlation function of the noise in the expression for the Brownian particle energy, we obtain now that

$$\langle V_t^2\rangle = \frac{\alpha_B}{2\alpha}\left[1 - e^{-2\alpha t}\right] \tag{47}$$

from which the long-time stationary value is indeed a non-zero finite value, in agreement with Statistical Mechanics

$$\langle V_t^2\rangle \longrightarrow \frac{\alpha_B}{2\alpha}, \quad t \to \infty. \tag{48}$$

Heuristic formulation. The previous manipulations led to interesting results but raise some questions:

1. Is it justified to handle F_t as a well-defined process?
2. Is it correct to integrate the initial equation as a deterministic one?

The main idea is not to handle the noise term directly but to manipulate the integration of the noise over a time interval, that is to consider

$$B_t = \int_0^t F(t')\,dt'. \tag{49}$$

This idea is simply making use of the smoothing properties of the integration operator. Then, in an heuristic approach, B_t is a Gaussian process with a mean function value $\langle B_t \rangle = 0$ and a correlation function expressed by

$$\langle B_t B_{t'} \rangle = \int_0^t du \int_0^{t'} \delta(u-v)\,dv = \min(t,t'). \tag{50}$$

However, we have seen in the preceeding section that this corresponds to the properties of the Wiener process and this suggests a possible identification of the integration of the white-noise term with the Wiener process, something which can be loosely written as

$$W_t = \int_0^t F(t')\,dt' \ (1) \quad \text{or} \quad F_t = \frac{dW_t}{dt} \ (2) \tag{51}$$

At first sight, these two formulations seem equivalent. Yet, since the trajectories of W_t are not differentiable, how can we give sense to these relations and, in particular, to the relation (2) above?

The answer to that question is that *we will not try to give sense to the form (2) but, using the smoothing properties of the integration operator, we will try to give a precise meaning to the form (1).* In more physical terms, we cannot give sense directly to the white-noise term but we may give sense to its effects over a small time interval. Translated back to the original equation we were considering, this means that we will not try to give a meaning to the equation

$$\frac{dX_t}{dt} = A(t, X_t) + B(t, X_t)\,\zeta_t \tag{52}$$

but to its integrated form based on the now-meaningful relation $\zeta_t\,dt = dW_t$

$$X_t = X_{t_0} + \int_{t_0}^t A(t_s, X_s)\,ds + \underbrace{\int_{t_0}^t B(s, X_s)\,dW_s}_{\text{stochastic integral}}. \tag{53}$$

As a short-hand notation, this last expression is usually written in an increment form:

$$dX_t = A(t, X_t)\,dt + B(t, X_t)\,dW_t. \tag{54}$$

The task ahead is now to define properly the stochastic integral.

2.6   Definition of the stochastic integal

The aim is to come up with a definition of the stochastic integral

$$I = \int_{t0}^{t} B(s, X_s)\,dW_s \tag{55}$$

or, using the more rigorous mathematical notations introduced at the beginning of this chapter, to define more generally the following integral

$$I(t, \omega) = \int_{t0}^{t} f(s, \omega)\,dW_s(\omega). \tag{56}$$

For each trajectory (that is for each ω), the first idea that comes to mind is to resort to a classical Riemann-Stieltjes definition of the integral, that is to propose $I(t, \omega) = \lim_{N \to \infty} I_N$ with

$$I_N = \sum_{i=1}^{N} f(\tau_i, \omega)\,(W_{i+1} - W_i) \text{ where } \tau_i \in [t_i; t_{i+1}]. \tag{57}$$

Such a definition would make sense if the discrete approximations I_N converge to a limit, independently of the choice of the intermediate times τ_i. In classical integration theory, this is ensured when the integrand function has bounded variation (Klebaner, 1998). However, as noted in the preceeding section, one property of the Wiener process is precisely that the sample paths are of unbounded variation in any interval! Consequently, classical Riemann-Stieltjes approach cannot be applied. Furthermore, it can even be shown that, if the limit of the discrete approximations I_N exist, this limit depends explicitly on the choice of the intermediate times τ_i...

The best way to illustrate this point is to consider the canonical example of stochastic integration, which is $I = \int_0^t W_s\,dW_s$. The classical Riemann-Stieltjes approximations are then given by

$$I_N = \sum_{i=1}^{N} W_{\tau_i}\,\left(W_{t_{i+1}} - W_{t_i}\right), \;\; \tau_i = \alpha\,t_i + (1-\alpha)\,t_{i+1} \;\;\; 0 \le \alpha \le 1. \tag{58}$$

It is straightforward to show that even the mean value of the stochastic integral is explicitly dependent on the value of the parameter α. Indeed, by using the properties of the Wiener process and in particular the form of the auto-correlation function, we have that

$$\langle I_N \rangle = \sum_{i=1}^{N} (\tau_i - t_i) = (1 - \alpha)\, t. \tag{59}$$

As a consequence, new approaches and new definitions of the stochastic integral are required. Two such definitions have been proposed, first by Ito (1941) with a rigorous mathematical basis, and later by Stratonovich (1965) with what can be loosely taken as a more physical standpoint. In the following, we will follow Ito's point of view which is now regarded as the standard definition and must always be assumed if nothing else is explicitly specified. Both definitions are developed for so-called *non-anticipating processes*: a stochastic process X_t is said to be non-anticipating if X_t is independent of the future of the Wiener process, which means that X_s is independent of $(W_t - W_s)$ for $s < t$.

Ito definition of the stochastic integral. For a non-anticipating process X_t, the stochastic integral is defined in the Ito sense as

$$\int_{t0}^{t} B(s, X_s)\, dW_s = \text{ms-} \lim_{N \to \infty} \sum_{i=1}^{N} B(t_i, X_{t_i}) \left(W_{t_{i+1}} - W_{t_i} \right) \tag{60}$$

where the limit must be understood as a limit in the mean-square sense (Arnold, 1974; Oksendal, 1995; Ottinger, 1996) and not any more as a convergence trajectory by trajectory. By comparison with the classical Riemann-Stieltjes expression, it is seen that the Ito definition consists in taking the value of the integrated function at the beginning of each time interval $\tau_i = t_i$. This means that the increments dW_i 'points into the future' while the function to integrate is frozen at the beginning of the small time interval. As a consequence of the properties of the Wiener increments and of the non-anticipating nature of the process, X_{t_i} depends only on $W_{t'}, t' \le t_i$ and therefore X_{t_i} and $\Delta W_i = W_{t_{i+1}} - W_{t_i}$ are independent. The non-symetrical choice in the Ito definition is not only essential for the mathematical justification of the existence of the limit but results in two fundamental properties of the integral:

$$\left\langle \int_{t0}^{t1} X_t dW_t \right\rangle = 0 \tag{61}$$

$$\left\langle \left(\int_{t0}^{t1} X_t dW_t \right) \left(\int_{t2}^{t3} Y_t dW_t \right) \right\rangle = \int_{t2}^{t1} \langle X_t Y_t \rangle dt \quad t_0 \le t_2 \le t_1 \le t_3. \tag{62}$$

As an example of the calculation of a stochastic integral with Ito definition, let us consider again the classical example of $I = \int_0^t W_s \, dW_s$ by developing the form of the discrete approximations

$$
\begin{aligned}
I_N &= \sum_{i=1}^{N} W_{t_i} \left(W_{t_{i+1}} - W_{t_i} \right) \\
&= \sum_{i=1}^{N} \frac{1}{2} \left\{ (W_{t_i} + W_{t_{i+1}} - W_{t_i})^2 - W_{t_i}^2 - (\Delta W_i)^2 \right\} \\
&= \sum_{i=1}^{N} \frac{1}{2} \left\{ W_{t_{i+1}}^2 - W_{t_i}^2 - (\Delta W_i)^2 \right\} \\
&= \frac{1}{2} \left(W_t^2 - W_{t_0}^2 \right) - \frac{1}{2} \underbrace{\sum_{i=1}^{N} (\Delta W_i)^2}_{A_N}
\end{aligned}
\tag{63}
$$

If the Wiener process was differentiable, then the variance of each increment entering the term A_N would be of order $(\Delta t)^2$ (considering a regular partition of the time interval $[0, t]$ with a constant time step Δt) and A_N would vanish as $\Delta t \to 0$. However, as emphasised before, this is not the case and the linear scaling of the variance of each increment with respect to Δt implies that A_N is not vanishing. Actually, it is possible to show that ms-$\lim_{N \to \infty} A_N = t$. Indeed, the random term A_N has a mean value

$$
\langle A_N \rangle = \sum_{i=1}^{N} \langle (\Delta W_i)^2 \rangle = \sum_{i=1}^{N} \Delta t = t
\tag{64}
$$

and by considering the variance of the difference, we get that

$$
\langle (A_N - t)^2 \rangle = \langle (\sum_{i=1}^{N} \left\{ (\Delta W_i)^2 - \Delta t \right\})^2 \rangle = \sum_{i=1}^{N} 2\Delta t^2 = 2\frac{t^2}{N} \to 0
\tag{65}
$$

which proves that, in the mean-square sense, the random term A_N is equal to the deterministic value t. Therefore, in the Ito sense, the integral is equal to $I = \frac{1}{2} \left(W_t^2 - W_{t_0}^2 \right) - \frac{1}{2} t$! If the first term on the right-hand side is the usual term we would expect from a direct application of the rules of classical calculus, the existence of the second term is an indication that the Ito definition implies new calculus rules. These new rules are referred to as the rules of 'stochastic calculus'.

Stochastic calculus. Altough tricky at first, the rules of stochastic calculus are actually easy to understand. Let us first formulate the rules of Ito calculus: If X_t is a stochastic process, solution of the following Stochastic Differential Equation (SDE) defined in the Ito sense

$$dX_t = A(t, X_t)\, dt + B(t, X_t)\, dW_t \qquad (66)$$

and if g a smooth-enough function (say $g \in \mathcal{C}^2$), then $Y_t = g(t, X_t)$ is the solution of the following SDE

$$dg(t, X_t) = \frac{\partial g}{\partial t}\, dt + A(t, X_t) \left(\frac{\partial g}{\partial x}\right)(t, X_t)\, dt + B(t, X_t) \left(\frac{\partial g}{\partial x}\right)(t, X_t)\, dW_t$$
$$+ \underbrace{\frac{1}{2} B^2(t, X_t) \left(\frac{\partial^2 g}{\partial x^2}\right)(t, X_t)\, dt}_{\text{new term}}. \qquad (67)$$

Thus, Ito calculus (or stochastic calculus) differ from classical calculus by the existence of an additional term which involves the second-order derivative. As illustrated by the example of the integral I (where it was equal to $-t/2$), this terms is crucial and corresponds to a (usually) non-zero deterministic term. Forgetting this extra term leads to drastic errors and false drifts with dramatic consequences in the manipulation of stochastic processes. The existence of the extra term entering Ito calculus compared to classical calculus is actually simple to understand: it stems from a Taylor-series development made to the second-order in dt since it must be remembered that $(dW_t)^2 = dt$ in a mean-square sense and, therefore (contrary to the rules of classical calculus with differentiable functions), second-order terms of the development can give contributions to the first order in dt. As a practical rule-of-the-thumb, it is often written that $dW_t \sim \sqrt{dt}$.

As an application of Ito calculus for the simple SDE $dX_t = A\, dt + B\, dW_t$, a direct calculation for X_t^2 shows that

$$d(X_t)^2 = (X_t + dX_t)^2 - X_t^2 = 2X_t\, dX_t + (dX_t)^2 = 2X_t\, dX_t + B^2\, dt \qquad (68)$$

where the last term on the right-hand side stems from the property of the Wiener increments.

We are now in a position to come back to our manipulation of the Langevin equation for Brownian motion that was the motivation for the introduction of white-noise terms

$$dV_t = -\alpha V_t\, dt + \alpha_B\, dW_t. \qquad (69)$$

With a proper formulation of the stochastic integral and a precise mathematical definition, we can now develop a well-defined calculation of the energy of the Brownian particles. Indeed, from Ito calculus

$$d(V_t)^2 = 2V_t\, dV_t + \alpha_B^2\, dt \Rightarrow d\langle V_t^2 \rangle = -2\alpha\langle V_t^2 \rangle\, dt + \alpha_B^2\, dt \tag{70}$$

which shows that statistical equilibrium is reached when

$$\alpha_B^2 = 2\alpha\langle V_t^2 \rangle = 2\alpha\frac{k_B\, T}{m_p}. \tag{71}$$

This results, which is also known as the fluctuation-dissipation theorem, shows that the noise and friction terms are related

$$dV_t = \underbrace{-\alpha V_t\, dt}_{\text{dissipative term}} + \underbrace{\sqrt{2\alpha k_B\, T/m_p}\, dW_t}_{\text{fluctuating term}} \tag{72}$$

3   Diffusion Stochastic Processes

3.1   Langevin and Fokker-Planck equations

Having properly defined SDEs, the correspondence between the trajectory point of view (which in physically-oriented presentations is often referred to as Langevin equations) and the PDF point of view (known as the Fokker-Planck equation) can be worked out.

Let us take X_t the solution of the SDE, $dX_t = A\, dt + B\, dW_t$, with $X(t_0) = x_0$. For any smooth function g, we consider $\langle g(X_t) \rangle$ and, since we are considering particles starting from a certain initial condition, this mean value can be expressed with the transitional pdf as

$$\langle g(X_t) \rangle = \int g(x)p(t, x \,|\, t_0, x_0)\, dx \tag{73}$$

from which it derives that

$$\frac{d}{dt}\langle g(X_t) \rangle = \int g(x)\frac{\partial p}{\partial t}(t, x \,|\, t_0, x_0)\, dx. \tag{74}$$

On the other hand, from the application of Ito calculus, we have that

$$d\langle g(X_t) \rangle = \langle A(t, X_t)(\frac{\partial g}{\partial x})(t, X_t) + \frac{1}{2}B^2(t, X_t)(\frac{\partial^2 g}{\partial x^2})(t, X_t) \rangle\, dt \tag{75}$$

since the mean term involving dW_t vanishes thanks to the nice properties of Ito definition of the stochastic integral. By collating the two expressions for $\langle dg(X_t)\rangle$, we end up with the identity

$$\int g(x)\frac{\partial p}{\partial t}(t,x\,|\,t_0,x_0)dx = \int \left\{A(t,x)\frac{\partial g}{\partial x} + \frac{1}{2}B^2(t,x)\frac{\partial^2 g}{\partial x^2}\right\}p(t,x\,|\,t_0,x_0)\,dx$$

$$= \int g(x)\left\{-\frac{\partial\,[\,A(t,x)\,p\,]}{\partial x} + \frac{1}{2}\frac{\partial^2\,[\,B^2(t,x)\,p\,]}{\partial x^2}\right\}dx. \tag{76}$$

Since this identity is valid for any smooth test function g, we obtain that, in a weak sense (or as a distribution), there is an equivalence between the trajectory point of view (the Langevin equations)

$$dX_t = A(t,\,X_t)\,dt + B(t,\,X_t)\,dW_t \quad \text{with } X(t_0) = x_0 \tag{77}$$

and the PDF point of view (the Fokker-Planck equation)

$$\begin{cases} \dfrac{\partial p}{\partial t} = -\dfrac{\partial[\,A(t,x)\,p\,]}{\partial x} + \dfrac{1}{2}\dfrac{\partial^2[\,B^2(t,x)\,p\,]}{\partial x^2} \\[2mm] p(t,x\,|\,t_0,x_0) = \delta(x - x_0) \quad \text{when } t \to t_0. \end{cases} \tag{78}$$

The coefficients A and B in the Langevin and Fokker-Planck equations are explained by the statistics of the conditional increments

$$\langle \Delta X | X(t) = x\rangle = A(t,x)\,\Delta t, \tag{79}$$

$$\langle (\Delta X)^2 | X(t) = x\rangle = B^2(t,x)\,\Delta t. \tag{80}$$

As a result and as displayed in Fig. 4, it is seen that these two coefficients represent two clear physical phenomena:

- The 'determistic' term, $A(t,x)$, governs the mean evolution of the conditional increments of the process: it is a **drift** term
- The 'random' term, $B(t,x)$, governs the spread of the conditional increments around its mean value: it is a **diffusion** term

It is worth clarifying an important point related to the Gaussian hypothesis and which, unfortunately, can lead to repeated confusion in the literature when the mathematical aspects have been skipped. From the properties of the Wiener process, it is clear that the increments ΔW_t over a time step Δt are indeed Gaussian random variables. As a consequence, there is a Gaussian hypothesis built in the Langevin equations. However, as clarified by the relations written above, only the **conditional increments** $\Delta X | (X(t) = x)$ of the process X_t are assumed to follow a Gaussian spread over the time interval Δt. In the very special case when the diffusion coefficient is constant

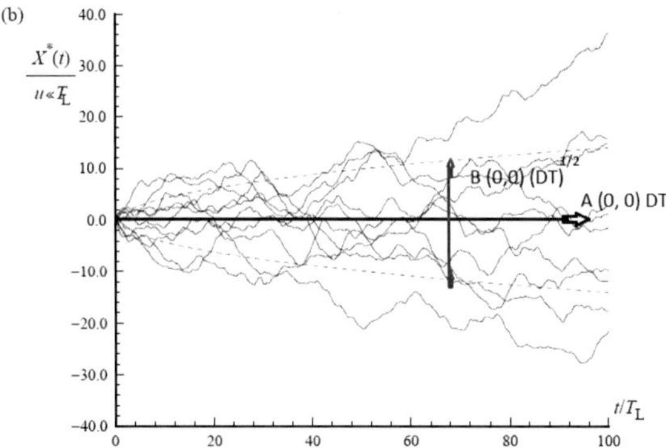

Figure 4. Some conditional trajectories of a stochastic diffusion process, or Langevin equations, illustrating the physical meaning of the drift and diffusion coefficients.

for example (as in the case of the original Langevin equation), then the process X_t itself becomes Gaussian. However, in the general case when $B(t,x)$ is not constant, the Gaussian hypothesis is only valid for the conditional increments and the resulting process X_t can (and usually does) deviate from Gaussianity!

The correspondence between Langevin and Fokker-Planck equations can easily been extended to the multi-dimensional case. For instance, we consider the SDEs written for a n-dimensional process $\mathbf{X}(t) = (X_1, \ldots, X_n)$

$$dX_i = A_i(t, \mathbf{X}(t))\, dt + B_{ij}(t, \mathbf{X}(t))\, dW_j \tag{81}$$

where $\mathbf{A} = (A_i)$ is the drift vector and $\mathbf{B} = (B_{ij})$ the diffusion matrix. Ito stochastic calculus for a process $g(t, \mathbf{X}_t)$ takes a similar form:

$$dg(t, \mathbf{X}_t) = \frac{\partial g}{\partial t}\, dt + \sum_{i=1}^{d} \frac{\partial g}{\partial x_i}(t, \mathbf{X}_t)\, dX_i(t)$$

$$+ \frac{1}{2} \sum_{i,j=1}^{d} \frac{\partial^2 g}{\partial x_i \partial x_j}(t, \mathbf{X}_t)(BB^T)_{ij}\, dt. \tag{82}$$

Following the same reasoning as in the one-dimensional case, it is straightforward to derive the multi-dimensional PDF equation (the Fokker-Planck equation)

$$\frac{\partial p}{\partial t} = -\frac{\partial [\, A_i(t, \mathbf{x})\, p\,]}{\partial x_i} + \frac{1}{2} \frac{\partial^2 [\, (BB^T)_{ij}(t, \mathbf{x})\, p\,]}{\partial x_i \partial x_j}. \tag{83}$$

3.2 Non-linear PDF equations and McKean diffusion processes

The mathematical theory is well-established when the drift and diffusion coefficients are functions of the value of the process: $A(t, x)$ and $B(t, x)$. This was the case presented in the previous subsection and it can be seen that the resulting Fokker-Planck equation is a linear equation with respect to $p(t, x)$. However, in many physical applications, we need to consider cases when the drift and diffusion coefficients become also functions of averages of the process: $A(t, x, \langle \mathcal{F}(X_t) \rangle)$ and $B(t, x, \langle \mathcal{G}(X_t) \rangle)$.

To provide a simple example we may wish to go for a standard Ornstein-Uhlenbeck process (Gardiner, 1990)

$$dX = -\frac{X}{T}\, dt + B(t, X)\, dW \tag{84}$$

to a generalised Ornstein-Uhlenbeck process such as:

$$dX = \underbrace{\left(\frac{d\langle X \rangle(t)}{dt} - \frac{X - \langle X \rangle(t)}{T} \right)}_{A(t, x, \langle X \rangle)} dt + B(t, X)\, dW \tag{85}$$

where the drift term depends on the average $\langle X \rangle(t) = \displaystyle\int x\, p(t, x)\, dx$ and, therefore, on the law (or the pdf) of the process.

The theory developed so far for the definition of the SDEs (the Langevin equations) as well as the derivation of the PDF equation (the Fokker-Planck equation) can be extended and remain valid but the Fokker-Planck equation, which can written as

$$\begin{cases} \dfrac{\partial p}{\partial t} = -\dfrac{\partial[\, A(t, x, \mathcal{H}_A(p))\, p\,]}{\partial x} + \dfrac{1}{2}\dfrac{\partial^2[\, B^2(t, x, \mathcal{H}_B(p))\, p\,]}{\partial x^2} \\ p(t, x \,|\, t_0, x_0) = \delta(x - x_0) \ \text{ when } \ t \to t_0. \end{cases} \tag{86}$$

to exhibit the dependence of the drift and diffusion coefficients with p, is now a non-linear equation with respect to the transitional pdf of the process. Correspondingly, the evolution equations for the trajectories of the process have the general form

$$dX_t = A(t, X_t, \langle \mathcal{F}(X_t) \rangle)\, dt + B(t, X_t, \langle \mathcal{G}(X_t) \rangle)\, dW. \tag{87}$$

In many physically-oriented texts, this distinction is hardly ever mentioned. However, it is worth being aware that this general case is in fact an extension of the well-established Langevin-Fokker-Planck framework. In the mathematical literature, these processes are referred to as *McKean diffusion processes* and in careful physical presentations (Ottinger, 1996) they are referred to as *processes with mean-field interactions*. In practical simulations and using classical Monte Carlo methods, these processes are approximated by *weakly interacting processes* and the evolution SDE for a trajectory labelled (i) is given by

$$dX_t^{(i)} = A(t, X_t^{(i)}, \frac{1}{n}\sum_{j=1}^{n} \mathcal{F}(X_t^{(j)}))\, dt + B(t, X_t^{(i)}, \frac{1}{n}\sum_{j=1}^{n} \mathcal{G}(X_t^{(j)}))\, dW \quad (88)$$

with $\mathrm{ms} - \lim_{n\to\infty} \frac{1}{n}\sum_{j=1}^{n} \mathcal{F}(X_t^{(j)}) = \langle \mathcal{F}(X_t)\rangle$.

3.3   General jump-diffusion processes

When discussing Markov processes, we have introduced and discussed separately the Poisson process, which leads to jump processes, and the Wiener process, which leads to diffusion processes. Such a distinction is helpful to study the specific characteristics of each subclass of Markov processes. However, these characteristics can be gathered in so-called jump-diffusion processes. Once again, these processes can be addressed from a trajectory or a PDF point of view. If we first consider the PDF point of view, then the extension of the Fokker-Planck equation to include jumps is known as the general Chapman-Kolmogorov equations and is discussed for example in Gardiner (1990, chap. 3.4). The drift, diffusion and jump amplitudes are defined from the transitional pdf $p(t + dt, y \,|\, t, x)$ by

$$\lim_{dt\to 0} \frac{1}{dt} p(t + dt, y \,|\, t, x) = W(y \,|\, t, x), \text{ for } |x - y| \geq \epsilon \quad (89)$$

$$\lim_{dt\to 0} \frac{1}{dt} \int_{|y-x|<\epsilon} (y - x) p(t + dt, y \,|\, t, x)\, dy = A(t, x), \quad (90)$$

$$\lim_{dt\to 0} \frac{1}{dt} \int_{|y-x|<\epsilon} (y - x)^2 p(t + dt, y \,|\, t, x)\, dy = B^2(t, x) \quad (91)$$

where the last two conditions are identical to the definitions of the drift and diffusion coefficients already introduced as the first and second-order moments of the conditional increment while the first condition defines the probablity of making a jump from state x to state y at time t. It can be shown that the general Chapman-Kolmogorov equation satisfied by the transitional pdf $p(t + dt, y \,|\, t, x)$ has the form

$$\frac{\partial p}{\partial t} = -\frac{\partial[\,A(t,x)\,p\,]}{\partial x} + \frac{1}{2}\frac{\partial^2[\,B^2(t,x)\,p\,]}{\partial x^2}$$
$$+ \int [\,W(x\,|\,t,y)p(t,y\,|\,t_0,x_0) - W(y\,|\,t,x)p(t,x\,|\,t_0,x_0)\,]\,dy. \quad (92)$$

The relation of this PDF equation with classical PDF equations is discussed in more details elsewhere (Minier and Peirano, 2001).

From the trajectory point of view, the SDEs for a jump-diffusion process are written as

$$dX_t = A(t,X_t)\,dt + B(t,X_t)\,dW_t + C(t,X_t)\,dN_t \quad (93)$$

where N_t a Poisson process with intensity λ and $C(t,X_t)$ is the amplitude of the jumps. The jumps contribute to the statistics of the conditional increments over a time step Δt

$$\langle \Delta X_t \,|\, X(t) = x \rangle = (A(t,x) + C(t,x)\lambda)\,\Delta t \quad (94)$$
$$\langle (\Delta X_t)^2 \,|\, X(t) = x \rangle = \left(B^2(t,x) + C^2(t,x)\lambda\right)\,\Delta t \quad (95)$$

The introduction of jumps is illustrated in Fig. 5 which displays sample paths of a pure diffusion process and in Fig. 6 which displays similar sample paths for a jump-diffusion process, revealing the discontinuous nature of the trajectories and their diffusive behaviour between successive jumps.

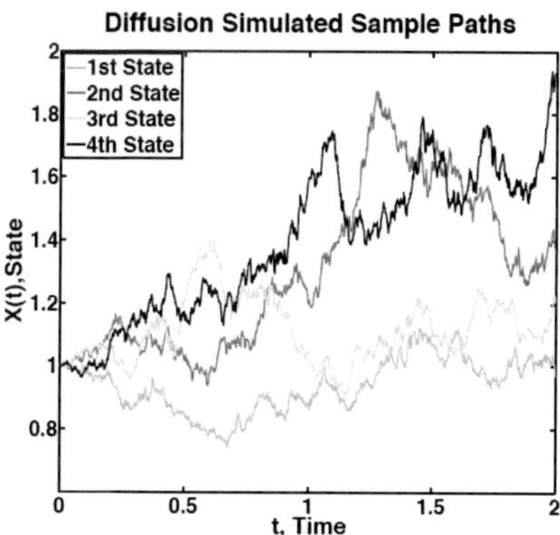

Figure 5. Four trajectories, or sample paths, of a diffusion process.

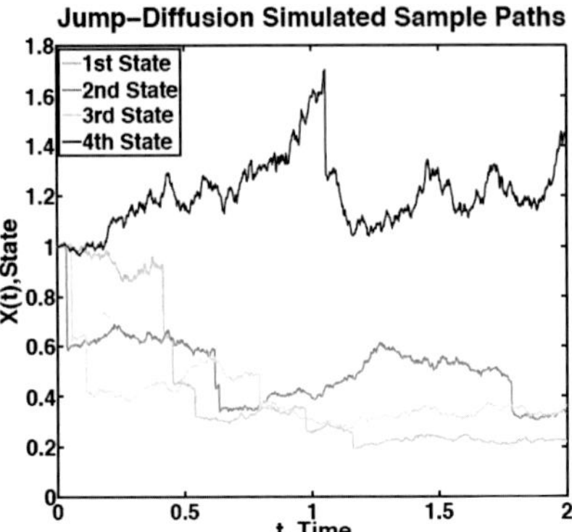

Figure 6. Four trajectories, or sample paths, of a jump-diffusion process.

Over an infinitesimal time increment dt, as explained above, the increments of a Poisson process can take only two possible values

$$\mathbb{P}[\,dN(t) = k\,] = (1 - \lambda dt)\,\delta_{k,0} + \lambda\,dt\,\delta_{k,1} \Rightarrow \langle (dN(t))^m \rangle = \lambda\,dt \quad \forall m. \quad (96)$$

Thus, the Poisson jumps contribute to any order in dt (while it is interesting to remember that the increments of the Wiener process contributes only to the second order in dt)!

The stochastic differential equations can be further generalised by considering random amplitudes for the jump, based on what is referred to as a compound Poisson process

$$dX_t = A(t, X_t)\,dt + B(t, X_t)\,dW_t + C(t, X_t, Q)\,dN_t \quad (97)$$

where Q is an independent random variable (in particular, independent of N_t). For such a general jump-diffusion process, the law of the random jumps is expressed by

$$W(y \mid X(t) = x) = \mathbb{P}[\,C = y | X(t) = x\,]\,\lambda \quad (98)$$

where the probability $\mathbb{P}$ is taken with respect to the law of the independent random variable Q.

4   Monte Carlo solutions of partial differential equations

4.1   Monte Carlo method for random variables

In this part, we briefly recall the key aspects of classical Monte Carlo methods, as applied to random variables. The basic idea is to use random numbers, or copies of a random variable, to estimate integrals and, in particular, statistics derived from this random variable. Indeed, if we consider a random variable X having a pdf $p(x)$, where x denotes the possible values taken by the random variable in its sample space, any statistics can be expressed as

$$I = \langle g(X) \rangle = \int_{\mathcal{D}} g(x)\, p(x)\, dx. \tag{99}$$

The Monte Carlo method consists in generating N identical and independent random variables, $(X_i)_{1 \leq i \leq N}$, with the same law $p(x)$ (basically N samples or copies of X) and to approximate the statistics by

$$I \approx I_N = \frac{1}{N} \sum_{i=1}^{N} g(X_i). \tag{100}$$

Once the approximation has been proposed, the two main questions are
  1. when does this algorithm converge?
  2. what is the precision or what is the rate of convergence?
It is worth pointing out that, in the above approximation, the statistics $\langle g(X) \rangle$ are numbers but the discrete sums I_N are actually random variables and should be rigorously noted as $I_N(\omega)$. Consequently, the meaning of the convergence expected in the two questions above must be carefully precised.

The positive answer to the first question is ensured by the Law of the large numbers which states that: let $(X_i)_{1 \leq i \leq N}$, N independent random variables with the same law $p(x)$ as a random variable X, and such that $\langle X^3 \rangle < \infty$, then, for almost every ω, we have that

$$\langle X \rangle = \lim_{N \to \infty} \frac{1}{N} \sum_{i=1}^{N} X_i(\omega). \tag{101}$$

This result is often referred to as the strong law of the large numbers since it is seen that convergence is to be understood here as the almost-sure mode of convergence (see the discussion on the different modes of convergence), therefore a strong mode of convergence.

To obtain the precision of the method, the error ϵ_N is defined as

$$\epsilon_N = \langle X \rangle - \frac{1}{N} \sum_{i=1}^{N} X_i(\omega) \tag{102}$$

and the answer to the second question is provided by the Central Limit Theorem (CLT) which states that: let $(X_i)_{1 \leq i \leq N}$, N independent random variables with the same law $p(x)$ as a random variable X, and such that $\sigma^2 = \langle X^2 \rangle - \langle X \rangle^2 < \infty$, then

$$\frac{\sqrt{N}}{\sigma} \epsilon_N \to \mathcal{N}(0,1) \tag{103}$$

where $\mathcal{N}(0,1)$ denotes the Gaussian distribution of zero mean and unit standard deviation. From the CLT, it is thus clear that the mode of convergence of the error ϵ_N is to be understood as a convergence in law, that is in a weak sense. From the CLT, we get an immediate estimation of the error

$$\lim_{N \to \infty} \mathbb{P}\left(\frac{\sigma}{\sqrt{N}} c_1 \leq \epsilon_N \leq \frac{\sigma}{\sqrt{N}} c_2 \right) = \int_{C_1}^{c_2} \frac{1}{\sqrt{2\pi}} e^{-x^2/2} \, dx \tag{104}$$

from which we obtain the well-known *confidence intervals*: any Monte Carlo estimation of the integral, I_N, has a 95%-chance to be found within the interval

$$\left[I_N - 2\frac{\sigma}{\sqrt{N}} \; ; \; I_N + 2\frac{\sigma}{\sqrt{N}} \right] \tag{105}$$

The rate of convergence of Monte Carlo estimations is slow ($\epsilon_N \sim N^{-1/2}$) but this rate does not depends on the regularity of the functions g and, more importantly, it does depend on the space dimension! Monte Carlo methods are therefore fairly general and, though not the best choices in some cases in low-dimensions, they are always applicable and remain at the moment the only real possibility for problems in high-dimensional spaces.

For our later purposes, the key point is that the standard Monte Carlo method corresponds to an estimation, in a weak sense, of the pdf of the random variable. Indeed, Monte Carlo approximations consist in writing that

$$\forall Q, \quad \langle Q(X) \rangle \approx \frac{1}{N} \sum_{i=1}^{N} Q(X_i). \tag{106}$$

By introducing the discrete pdf $p_N(x)$ as a sum of dirac masses centered on the N samples

$$p_N(x) = \frac{1}{N} \sum_{i=1}^{N} \delta(x - X_i) \tag{107}$$

we have therefore that

$$\forall Q, \quad \int Q(x)p(x)\,dx \approx \int Q(x)p_N(x)\,dx \tag{108}$$

from which it derives that $p(x) \simeq p_N(x)$ in the weak sense (or as a distribution).

4.2 Dynamical Monte Carlo methods for stochastic processes

The weak approximation of the pdf by a sum of dirac distributions is the central point for practical simulations of stochastic processes, which corresponds to what can be referred to as *dynamical Monte Carlo methods*. This method is a direct extension of the standard Monte Carlo approach for random variables and can be used for any Markov stochastic processes. However, for the sake of clarity and since we have been mostly concerned with diffusion stochastic processes, we will present and illustrate the dynamical Monte Carlo method for these processes.

As detailed in the preceeding sections, a stochastic diffusion process can be described by two equivalent points of view: the PDF point of view and the trajectory point of view. The PDF point of view is acting in sample-space through the Fokker-Planck equation and the trajectory point of view is acting in physical space through the Langevin equations. The equivalence (in a weak sense) between the two points of view can be sketched as follows:

$$\underbrace{\begin{cases} p(0,\mathbf{x}) = p_0(\mathbf{x}_0) \text{ at } t = 0 \\ \dfrac{\partial p(t,\mathbf{x})}{\partial t} = \mathcal{L}_t^*[\,p(t,\mathbf{x})\,] \end{cases}}_{\textit{sample space}} \Leftrightarrow \underbrace{\begin{cases} \mathbf{X}(0) = \mathbf{X}_0 \text{ at } t = 0 \\ dX_i(t) = A_i(t,\mathbf{X}(t))\,dt + B_{ij}(t,\mathbf{X}(t))\,dW_j \end{cases}}_{\textit{physical space}}$$

where $\mathcal{L}_t^*$ stands for the operator acting in phase-space on $p(t,\mathbf{x})$

$$\mathcal{L}_t^* = -\frac{\partial\,[A_i(t,\mathbf{x})\,.]}{\partial x_i} + \frac{\partial^2\,[(BB^T)_{ij}(t,\mathbf{x})\,.]}{\partial x_i \partial x_j}$$

as an example of a forward Kolmogorov equation. It was stressed above that for a Markov process, the key function is the transitional pdf $p(t,\mathbf{x}\,|\,t_0,\mathbf{x}_0)$. However, for the sake of clarity and in order to avoid the more cumbersome notations of the transitional pdf, we have considered here the one-time pdf $p(t,\mathbf{x})$. Actually, the equations satisfied by the transitional pdf and by the one-time are identical since the latter one is derived from the former one by integration over all possible initial conditions. Therefore, the presentation that follows will be developed in terms of $p(t,\mathbf{x})$, as a sort of short-cut, but remains valid and applicable for the transitional pdf itself.

The dynamical Monte Carlo method is illustrated by the sketch below. At the initial time $(t = 0)$, we have a random variable and we can thus apply the standard Monte Carlo method and introduce a (large) number of samples, from which the initial pdf is approximated by a sum of Dirac masses. As time evolves $(t > 0)$, the initial random variable becomes a stochastic process and the dynamical aspects can be addressed from two points of view. As illustrated by the left part of the sketch, one could solve the evolution equation in sample space, that is the Fokker-Planck equation for our example of a stochastic diffusion process. This means solving one equation but in a space which can have a very high dimension (indeed, it will be seen that typical PDF models in two-phase flows involves a state-vector having at least 9 dimensions).... This road becomes quickly impossible for numerical solutions unless we limit ourselves to very small dimensional spaces. However, one could address the same evolution problem following the second point of view which is illustrated on the right part of the sketch. This amounts to tracking in time the evolution of the N initial samples or, in other terms, to simulating the trajectories of these N samples or 'particles'. At a later time t, we have thus N updated samples (written as $\mathbf{X}^{(k)}(t)$) and we can apply the standard Monte Carlo method at that time, using these samples, to obtain directly a weak approximation of the pdf $p(t, x)$.

$$
\overbrace{}^{\textit{sample space}} \qquad \overbrace{}^{\textit{physical space}}
$$

$$
\left\{ \begin{array}{l} \mathbf{X}(0)\,\text{r.v with a density} \\[2ex] p(0, \mathbf{x}) = p_0(\mathbf{x}) \end{array} \right. \quad \underset{(t=0)}{\Longleftrightarrow} \quad \left\{ \begin{array}{l} N\,\text{samples}: \mathbf{X}^{(k)}(0),\ k = 1, \ldots N \\[2ex] p(0, \mathbf{x}) \approx \dfrac{1}{N} \sum_{k=1}^{N} \delta(\mathbf{x} - \mathbf{X}^{(k)}(0)) \end{array} \right.
$$

$$
\Downarrow (t > 0)
$$

$$
\begin{array}{cc}
\text{solution of the PDE} & \text{advance of the SDEs} \\
(\text{1 equation in } (t, \mathbf{x}) \text{ in a } N_d\text{-space}) & (N \text{ equations in } t, \text{ in 3D-space})
\end{array}
$$

$$
\left\{ \begin{array}{l} \mathbf{X}(t)\,\text{r.v with a density} \\[2ex] p(t, \mathbf{x}) \end{array} \right. \quad \Longleftrightarrow \quad \left\{ \begin{array}{l} N\,\text{samples}: \mathbf{X}^{(k)}(t),\ k = 1, \ldots N \\[2ex] p(t, \mathbf{x}) \approx \dfrac{1}{N} \sum_{k=1}^{N} \delta(\mathbf{x} - \mathbf{X}^{(k)}(t)) \end{array} \right.
$$

In that sense, it can be seen that the dynamical Monte Carlo method corresponds to making an approximation for the solution of a PDF equation, which is a partial differential equation in sample-space, without actually 'solving' the PDF equation. In practise, this dynamical Monte Carlo method is often referred to as *particle stochastic method*. Since the cost is primarily related to the number of 'stochastic particles' used (the number N of samples) rather than the dimensional of the sample space, the dynamical Monte Carlo is a universal method, particularly attractive for high-dimensions, and thus more or less the only general approach available if one is to make a real PDF simulation. In this approach, the issue is therefore to be able to integrate the values along the trajectories of the process, based on specific numerical schemes devoted to SDEs (Peirano et al., 2006).

4.3   Probabilistic solutions of Partial Differential Equations

The relation between Langevin equations (SDEs) and Fokker-Planck equation (PDE) is the source of many interesting solutions of Partial Differential Equations. For example, the solution of the evolution problem

$$\begin{cases} \dfrac{\partial u}{\partial t} = -\dfrac{\partial[\,A(t,x)\,u\,]}{\partial x} + \dfrac{1}{2}\dfrac{\partial^2[\,B^2(t,x)\,u\,]}{\partial x^2} \\ u(0,x) = h(x) \;\; \text{when} \;\; t = 0, \end{cases} \tag{109}$$

is built from the transitional pdf of the diffusion process X_t, which has A and B as drift and diffusion coefficients, and is given by

$$u(t,x) = \int p(t,x\,|\,t_0,x_0)h(x_0)\,dx_0 \tag{110}$$

since $p(t,x\,|\,t_0,x_0)$ is the solution of the corresponding forward Kolmogorov equation (with a Dirac function as an initial condition). Therefore, the solution can be expressed as

$$u(t,x) = \langle\, h(X(0))\,|\,X_t = x\,\rangle \tag{111}$$

where the expectation is taken with respect to the diffusive trajectories, which are solution of the SDE

$$dX(t) = A(t,x)\,dt + B(t,x)\,dW_t$$

and *arriving* at location x at time t.

It may be interesting to illustrate this result with a discrete (or particle) formulation by considering that we assign a given value, say a 'scalar' variable ϕ, to each stochastic particle following the diffusive evolution. The corresponding state-vector is thus extended to include both location and scalar (X,ϕ). The scalar is given an initial value equal to the local value of the field h at the initial location $X(0)$ and remains constant along each particle trajectory. The particle SDEs are therefore expressed by

$$
\begin{cases}
X(0) = X_0 \text{ and } \Phi(0) = h(X_0) \text{ at } t = 0 \\
dX(t) = A(t, X(t))\, dt + B(t, X(t))\, dW \\
d\Phi(t) = 0
\end{cases}
\tag{112}
$$

From the dynamical Monte Carlo method, the discrete joint pdf is

$$
p_N(t, x, \phi) = \frac{1}{N} \sum_{k=1}^{N} \delta(x - X^{(k)}(t))\, \delta(\phi - \Phi^{(k)}(t))
\tag{113}
$$

We define the field $u(t, x) = \langle \phi \,|\, X_t = x \rangle$, which represents the 'averaged' value of the 'scalar' for particles arriving at x at time t (this is an example of an Eulerian statistics defined as a conditional Lagrangian statistics). In the Monte Carlo simulation, this field is estimated by

$$
u(t, x) = \frac{1}{p(t, x)} \int \phi\, p(t, x, \phi)\, d\phi \simeq \frac{1}{p_N(t, x)}\, \frac{1}{N} \sum_{k=1}^{N} \delta(x - X^{(k)}(t))\, \Phi^{(k)}(t)
\tag{114}
$$

The dirac functions are 'smoothed' around x: we consider a small volume $\delta \mathcal{V}_x)$ around the position x which contains N_x particles in it. Then, we approximate the dirac functions with

$$
\delta(y - x) \simeq \frac{1}{\delta \mathcal{V}_x} \mathbb{1}_{(y \in \delta \mathcal{V}_x)}, \quad p_N(t, x) \simeq \frac{N_x}{N\, \delta \mathcal{V}_x}
\tag{115}
$$

which gives for the local value of the field $u(t, x)$

$$
u(t, x) \simeq \frac{1}{p_N(t, x)}\, \frac{1}{N\, \delta \mathcal{V}_x} \sum_{k=1}^{N_x} \Phi^{(k)}(t) = \frac{1}{N_x} \sum_{k=1}^{N_x} \Phi^{(k)}(t)
\tag{116}
$$

Thus, we retrieve naturally the Monte Carlo estimation of the solution of the PDE since

$$
u(t, x) = \frac{1}{N_x} \sum_{k=1}^{N_x} h(X(0)) \simeq \langle\, h(X(0)) \,|\, X_t = x \,\rangle.
\tag{117}
$$

The backward point of view can also be applicable. Indeed, if we consider the following partial differential equation with an end condition

$$
\begin{cases}
\dfrac{\partial u}{\partial s} + A(s, x)\, \dfrac{\partial u}{\partial x} + \dfrac{1}{2} B^2(s, x)\, \dfrac{\partial^2 u}{\partial x^2} = r(s, x)\, u(s, x) \\[2mm]
u(T, x) = g(x) \quad \text{when} \quad s = T
\end{cases}
\tag{118}
$$

then the solution can be built from the transitional pdf of the diffusion process X_t (using the Kolmogorov backward equation) and is expressed by

$$u(s,x) = \int g(y) \underbrace{e^{-\int_s^T r(X(t'),X(t'))dt'}}_{w(X(T))} p(T,y \mid s,x)\, dy. \qquad (119)$$

The solution is thus given by

$$u(s,x) = \langle\, w(X(T))\, g(X(T)) \mid X_s = x \,\rangle \qquad (120)$$

where the expectation is taken now with respect to trajectories *leaving* the location x at time s. In the above formula, it is seen that the function $w(X(T))$ plays the role of a statistical weight attached to each diffusive trajectory (similar therefore to a sort of death-birth process along each trajectory). In the case when $r = 0$, the statistical weights remain equal to 1 for all particle trajectory and we retrieve a formula which appears as the 'adjoint' of the one developed above for the initial value problem. The expression in Eq. (119) played an historical role in the development of stochastic models and of path-integral formulations: this is the *celebrated Feynman-Kac formula*!

5   Summary and conclusion

Many applications in Fluid Mechanics (turbulence, particle transport, dispersion, ...) involve processes with continuous changes and stochastic diffusion processes can therefore be interesting modelling tools, while jump processes are natural candidates to model particle collisions for example.

It has been recalled or emphasised that stochastic diffusion processes can be addressed from two equivalent points of view: the PDF point of view and the trajectory point of view. From the trajectory point of view, the characterisation of the process is obtained with stochastic differential equations, referred to as Langevin equations

$$dX_t = A(t,X_t)\, dt + B(t,X_t)\, dW_t$$

which require a proper definition of the stochastic integral. From the PDF point of view, the evolution in sample-space is given by the Fokker-Planck equation. Both points of views are meaningful only for Markov processes since the transitional pdf is indeed the key function from which all information about the stochastic process can be retrieved.

In order to understand and manipulate properly stochastic diffusion processes, a sound knowledge of the mathematical definition of the stochastic integral and, in particular of Ito definition, is a mandatory step. Such a step not only allows researchers to be free from any misconceptions (whereas even distinguished physicists fell into the poor belief that Ito definition was 'a mathematical vagary'...) but, more importantly, will avoid them to induce spurious drifts by flawed mathematical calculus and also potentially inconsistent numerical schemes.

Bibliography

L. Arnold, *Stochastic Differential Equations, theory and applications*, John Wiley & Sons, New-York, 1974.

I. Karatzas and S. E. Shreve, *Brownian motion and stochastic calculus*, Springer Verlag, Berlin, 2cd edition, 1991.

B. Oksendal, *Stochastic Differential Equations, an introduction with applications*, Springer Verlag, Berlin, 5th edition, 2003.

F. C. Klebaner, *Introduction to stochastic calculus with applications*, Imperial Press College, London, 1998.

C. W. Gardiner, *Handbook of stochastic methods for Physics, Chemistry and the Natural Sciences*, Springer Verlag, Berlin, 2th edition, 1990.

H. C. Ottinger, *Stochastic processes in polymeric fluids*, Springer Verlag, Berlin, 1996.

N. G. Van Kampen, *Stochastic processes in Physics and Chemistry*, Elsevier, 1992.

S. B. Pope, *Turbulent flows*, Cambridge University Press, Cambridge, 2000.

S. B. Pope, *PDF Methods for Turbulent Reactive Flows*, Prog. Energ. Comb. Sci., Vol. 11, pages 119-192, 1985.

J.-P. Minier and E. Peirano, *The PDF approach to turbulent polydispersed two-phase flows*, Physics Reports, Vol. 352, N. 1-3, pages 1-214, 2001.

E. Peirano, S. Chibbaro, J. Pozorski and J.-P. Minier, *Mean-field/PDF numerical approach for polydispersed turbulent two-phase flows*, Prog. Energ. Comb. Sci., Vol. 32, pages 315-371, 2006.

Stochastic modelling of polydisperse turbulent two-phase flows

Sergio Chibbaro [*‡] and Jean-Pierre Minier [†]

[*] Institut D'Alembert University Pierre et Marie Curie, 4, place jussieu 75252 Paris Cedex 05,
[‡] CNRS UMR 7190, 4, place jussieu 75252 Paris Cedex 05
[†] Electricité de France, Div. R&D, MFEE, 6 Quai Watier, 78400 Chatou, France

1 Introduction

In this chapter we focus on the Lagrangian stochastic approach to turbulent polydispersed two-phase flows. This chapter has several objectives. The first important objective is to use this interesting and relevant physical subject to apply the mathematical techniques presented in the first chapter. It is important to understand how the stochastic approach actually works and what are the main issues related to it. The second objective is to offer the reader the possibility to become more familiar with particle-laden flows, a sub-field of fluid mechanics which is quite fascinating and important in many industrial and environmental applications. Third, this chapter offers a comprehensive but concise description of the whole formalism needed to develop the stochastic approach to turbulent polydispersed flows. While Lagrangian stochastic models have been put forward since the sixties for single-phase flows (Lundgren, 1967) and applied with success to reactive flows since the seventies (Pope, 1985, 1994), their development and diffusion for polydispersed flows is much more recent (Minier and Pozorski, 1999; Minier and Peirano, 2001). In this sense, the formalism generalises the reactive flow one. Finally, the study of a typical industrial application is shown. This test-case helps to clarify that stochastic models can be used to investigate realistic phenomena (they are computationally performing), and, at the same time, give satisfactory answers in complex problems, for which less refined approaches like two-fluid models are not able to give acceptable results.

2   Basic Concepts

2.1   Basic Equations

Dispersed two-phase flows are met when a continuous phase (a gas or a liquid) carries discrete particles (solid particles, droplets, bubbles, ...).

The basic equations of these flows are given by the Navier-Stokes equations together with the elementary behaviour of a single particle. The continuity and the Navier-Stokes equations where the different fields are density $\rho_f(t, \mathbf{x})$, pressure $P(t, \mathbf{x})$ and velocity $\mathbf{U}_f(t, \mathbf{x})$,

$$\frac{\partial U_{f,j}}{\partial x_j} = 0, \tag{1a}$$

$$\frac{\partial U_{f,i}}{\partial t} + U_{f,j}\frac{\partial U_{f,i}}{\partial x_j} = -\frac{1}{\rho_f}\frac{\partial P}{\partial x_i} + \nu\frac{\partial^2 U_{f,i}}{\partial x_j^2}, \tag{1b}$$

The equations for particles are less well-founded than Navier-Stokes equations for fluid particles and remain a subject of current research. For small particle-based Reynolds numbers Re_p (whose definition is specified below) and particle diameters that are of the same order of magnitude as the Kolmogorov length scale, a general form of the particle momentum equation has been proposed (Gatignol, 1983; Maxey and Riley, 1983). For the case considered of heavy particles, the equations of motion can be generalised to large Reynolds number and reduce to :

$$\frac{d\mathbf{x}_p}{dt} = \mathbf{U}_p, \tag{2a}$$

$$\frac{d\mathbf{U}_p}{dt} = \frac{1}{\tau_p}(\mathbf{U}_s - \mathbf{U}_p) + \mathbf{g}, \tag{2b}$$

where $\mathbf{U}_s = \mathbf{U}(\mathbf{x}_p(t), t)$ is the fluid velocity seen, *i.e.* the fluid velocity sampled along the particle trajectory $\mathbf{x}_p(t)$, not to be confused with the fluid velocity $\mathbf{U}_f = \mathbf{U}(\mathbf{x}_f(t), t)$ denoted with the subscript f. The particle relaxation time is defined as

$$\tau_p = \frac{\rho_p}{\rho_f}\frac{4d_p}{3C_D|\mathbf{U_r}|}, \tag{3}$$

where the local instantaneous relative velocity is $\mathbf{U_r} = \mathbf{U_s} - \mathbf{U_p}$ and the drag coefficient C_D is a non-linear function of the particle-based Reynolds number, $Re_p = d_p|\mathbf{U_r}|/\nu_f$, which means that C_D is a complicated function of the particle diameter d_p, (Clift et al., 1978). For example, a very often retained empirical form for the drag coefficient is

$$
C_D = \begin{cases} \dfrac{24}{Re_p}\left[\,1 + 0.15 Re_p^{0.687}\,\right] & \text{if } Re_p \leq 1000, \\[2mm] 0.44 & \text{if } Re_p \geq 1000. \end{cases}
\tag{4}
$$

In this chapter, we choose to limit ourselves to the case of dilute particle-laden flows, for the sake of clarity. However, it is worth underling that general Lagrangian stochastic approach and the formalism developped afterwards represent a framework of general scope and can be applied to more general situations. On the other hand, the modelling details are dependent on the specific physical phenomena which are considered and on the objectives.

Given the equations, an "exact approach" (in the spirit of DNS) is possible (Boivin et al., 1998; Marchioli and Soldati, 2002; Toschi and Bodenschatz, 2009), but in practice, the exact equations of motion are not of great help. Indeed, in the case of a large number of particles and of turbulent flows at high Reynolds numbers, the number of degrees of freedom is huge and a contracted probabilistic description is needed.

The presence of non-linear terms in the equations (2), leads to the known problem of hierarchy in the mean field equations, as in single-phase turbulence, so that an eventual Eulerian averaged approach to the problem would require a closure at this level. As explained by Pope (1994) for the turbulent reactive case (which is conceptually analogous to polydispersed turbulent two-phase flows case), a Lagrangian stochastic or PDF approach to the problem overcomes this issue with a general exact treatment of the non-linear terms. Moreover, since in the case of polydispersed two-phase flows the exact mean equations are not a priori known, mean field level closure appears often unsatisfactory.

2.2   Brownian Motion

Let us now turn our attention to a remarkable phenomenon, that is Brownian motion, which is quite useful to introduce the main themes of stochastic modelling, other than being a paradigmatic success of modern statistical physics. Observed on a microscope, pollen suspended in a glass of water moves erratically and incessantly, although the water appears to be still, and no work is done on pollen particles, to balance the energy dissipated by the viscosity of the fluid. This phenomenon was named after Robert Brown, the botanist who first tried to explain it as a form of life, which seemed to animate pollen particles suspended in a fluid (Brown, 1828). Among many other issues, the Brownian motion constitutes the ultimate evidence of the existence of atoms and led to the determination of the Boltzmann constant k_B in macroscopic experiments.

Why doesn't the motion of the pollen rapidly stops? The equation for the velocity of one spherical particle of mass m and radius R, subjected to no other forces than that exerted by the viscosity η of the liquid,

$$\frac{d\mathbf{v}}{dt} + \frac{6\pi R\eta}{m}\mathbf{v} = 0 \, , \tag{5}$$

predicts the exponentially decaying behaviour

$$\mathbf{v}(t) = \mathbf{v}(0)\exp(-t/\tau) \, , \quad \text{with} \quad \tau = \frac{m}{6\pi R\eta} \, , \tag{6}$$

where $\mathbf{v}(0)$ is the initial velocity, $\mathbf{v}(t)$ the velocity at a subsequent time t and τ is a characteristic time depending on the properties of both water and pollen. For pollen of radius $R \sim 10^{-4}$ metres, one obtains $\tau \sim 10^{-4}$ seconds, which means that $\mathbf{v}(t)$ should practically vanish in a few milliseconds.

The observation made in the second half of the nineteenth century, that the velocity of pollen increases with temperature, while it decreases with the pollen size and with the fluid viscosity, suggested that the kinetic theory of gases could explain the phenomenon. At the beginning of the twentieth century, Einstein and Smoluchowski proposed a theory, which Langevin put forward in modern terms as follows: the motion of pollen is determined by two forces:

- the deterministic viscous force obtained from Stokes law,
- a stochastic force due to the collisions with water molecules, which bears no memory of events occurring at different times.

This implies that Eq.(5) should be modified as:

$$\frac{d\mathbf{v}}{dt} + \frac{6\pi R\eta}{m}\mathbf{v} = f_R(t) \, , \tag{7}$$

where f_R is a random force representing the action of the water molecules on the pollen grains which, in accord with kinetic theory, is more energetic at higher temperatures. The randomness is reflected in the lack of correlations between the action of the water molecules at different time instants and is justified by the vast separation of the temporal scales concerning microscopic impacts (of order $10^{-12} - 10^{-11}$ seconds) and those concerning the macroscopic viscous damping (of order $10^{-5} - 10^{-4}$ seconds). Furthermore the average of the random force is required to vanish so that: the average work vanishes, there is no loss of energy in time, and pollen may then persist in its motion forever.

The result of this theory is the Einstein-Smoluchowski diffusion law (Castiglione et al., 2008; Boffetta and Vulpiani, 2012),

$$\langle \mathbf{x}^2(t) \rangle \simeq 6Dt , \quad \text{where } D = \frac{k_B T}{6\pi\eta\mathcal{R}} , \tag{8}$$

in which $\mathbf{x}(t)$ is the displacement of a pollen grain at time t from its initial position, $\mathbf{x}(0)$, and T is the common temperature of water and pollen. The constant D is known as the *diffusion coefficient*. To solve the problem, Einstein and Smoluchowski used a bold hypothesis: the equilibrium between fluid molecules and the pollen grain, which allowed the use of equipartition theorem (Castiglione et al., 2008).

Equation (8) turned out to be extremely important, since it connected easily measurable macroscopic quantities, such as $\langle \mathbf{x}^2(t) \rangle$, with Avogadro's number, which could at last be estimated.[1] Interestingly, Einstein had correctly anticipated that relations concerning fluctuations could be used to investigate the microscopic realm by means of macroscopic observations (Einstein, 1956a).

The agreement between theory and experiments was demonstrated by Perrin only a few years later, a result which convinced everyone that atoms could indeed be "counted" and "measured", hence, that they had to exist. To prove (or disprove) the existence of atoms was indeed Einstein's purpose (Einstein, 1956b).

This means that qualitatively different representations of matter, like thermodynamics and kinetic theory, are required to describe observations which take place on the corresponding hugely different scales.

Observing Eq.(7), one finds that there are two different limiting situations, involving the mass of pollen particles. The first is the limit of large mass, which makes the effect of the molecular impacts negligible, compared to that of the fluid viscosity. The second is that of small pollen mass, which makes dominant the effect of the molecular impacts, with respect to the viscous forces.

In such a way, the Brownian motion shows in a clear manner that, when a large separation of scale is present, the effect of fast scales can be correctly modelled by a stochastic term acting on large scales.

Hence, the theory of Brownian motion unveils one level of description, the mesoscopic level, which is as hard to connect to the microscopic and the macroscopic levels. The ingenuity of the Einstein-Smoluchowsky theory lies in its ability to identify three separate scales concerning objects in thermodynamic equilibrium, and to link them, by allowing microscopic and

---

[1] At that time, N_A was but a parameter of the atomic theory, whose value was unknown.

macroscopic forces act at once on pollen, and by using kinetic theory to determine the viscosity of the fluid. The phenomena which can be observed at the different scales, obviously coexist, but correspond to such widely separated scales that completely different kinds of description are required to understand them. This is what makes Brownian motion possible: the separation of scales. The mass of pollen is so much larger than the mass of liquid molecules, that the relaxation time τ of Eq.(6) is much larger than the molecular collision times and, at the same time, the mass of pollen is so much lighter than any macroscopic object, that the energy exchanged in molecules-pollen impacts suffices for thermal equilibrium to be established in times much shorter than τ.

2.3   State-vector

In the Langevin treatment of the Brownian motion (Langevin, 1908), the relevant variables to describe the whole physics are the velocity and the position: $\mathbf{x}, \mathbf{v}$. These two variables define therefore the state-vector of the system. Of course other choices can be possible, for instance the sole position as it was done by Einstein, or adding other variables like acceleration. It appears that a hierarchy between state-vectors naturally arises with regard to the information content desired in the physical approach (Minier and Peirano, 2001): the more information is detailed, the larger the number of variables possibly contained in the state vector. Resorting to Brownian motion, if we restrict ourselves to follow the position of the particle, the state vector is $\mathbf{Z}(t) = (\mathbf{X}(t))$, the particle velocity is an external variable and the pdf equation for $p(t, \mathbf{y})$ is unclosed

$$\frac{\partial p(t, \mathbf{y})}{\partial t} + \frac{\partial}{\partial \mathbf{y}} \left(\langle \mathbf{U} \,|\, \mathbf{y} \rangle \, p(t, \mathbf{y}) \right) = 0. \tag{9}$$

To obtain a closed model, the effect of the particle velocity has to be replaced by a model

$$\frac{dX^+(t)}{dt} = U^+(t) \Longrightarrow \frac{dX(t)}{dt} = F[t, X(t)] \tag{10}$$

where the superscript $^+$ denotes the exact equation and $F[t, X(t)]$ represents a functional of the position $X(t)$. If the functional F is deterministic we end up with a reduced Liouville equation. If this first picture is believed to be too crude, one can include the velocity of the particle in the state vector that becomes then $\mathbf{Z}(t) = (\mathbf{X}(t), \mathbf{U}(t))$ (Langevin's point of view). In this picture, the particle acceleration $\mathbf{A}(t)$ is an external variable and the corresponding pdf equation for $p(t, \mathbf{y}, \mathbf{V})$ is unclosed

$$\frac{\partial p(t, \mathbf{y}, \mathbf{V})}{\partial t} + \frac{\partial \left(V_i p(t, \mathbf{y}, \mathbf{V})\right)}{\partial y_i} + \frac{\partial}{\partial V_i} \left(\langle \mathbf{A} \,|\, \mathbf{y}, \mathbf{V} \rangle \, p(t, \mathbf{y}, \mathbf{V})\right) = 0. \qquad (11)$$

To obtain a closed form, the acceleration has to be eliminated or replaced by a model

$$\begin{cases} \dfrac{dX^+(t)}{dt} = U^+(t) \\[2mm] \dfrac{dU^+(t)}{dt} = A^+(t) \end{cases} \implies \begin{cases} \dfrac{dX(t)}{dt} = U(t) \\[2mm] \dfrac{dU(t)}{dt} = F[t, X(t), U(t)]. \end{cases}$$

It is thus clear that the second description encompasses the first one. It contains more information and in physical terms corresponds to a description performed with a finer resolution. From a modelling point of view the task is also different depending upon the choice of the one-particle state vector. In the first case (Einstein's point of view), one has to model particle velocities. In the second case (Langevin's point of view) one has to model particle accelerations.

From the above example, a general picture emerges. We consider a one-particle reduced description but with many internal degrees of freedom, *i.e* $\mathbf{Z} = (Z_1, Z_2, \ldots, Z_p, \ldots)$. If the time rate of change of the particle degrees of freedom has the following form

$$\frac{dZ_1}{dt} = g(t, Z_1, Z_2), \qquad (12\text{a})$$

$$\frac{dZ_2}{dt} = g(t, Z_1, Z_2, Z_3), \qquad (12\text{b})$$

$$\vdots \qquad (12\text{c})$$

$$\frac{dZ_p}{dt} = g(t, Z_1, \ldots, Z_p, Z_{p+1}), \qquad (12\text{d})$$

$$\vdots \qquad (12\text{e})$$

and if the chosen one-particle reduced state vector contains only a limited number of degrees of freedom, say p, $Z^r = (Z_1, \ldots, Z_p)$ then the corresponding pdf equation for $p^r(t, z_1, z_2, \ldots, z_p)$ is generally unclosed since it involves an external variable, namely Z_{p+1}

$$\frac{\partial p^r}{\partial t} + \frac{\partial \left(g(t, z_1, z_2) \, p^r\right)}{\partial z_1} + \ldots$$
$$+ \frac{\partial \left(g(t, z_1, \ldots, z_p) \, p^r\right)}{\partial z_{p-1}} + \frac{\partial \left(\langle g(t, Z_1, \ldots, Z_p, Z_{p+1}) \, | \, Z^r = z^r \rangle \, p^r\right)}{\partial z_p} = 0.$$

(13)

To obtain a closed model, the external variable Z_{p+1} must be expressed as a function of the variables contained in the chosen state vector, and the equations for the modelled system have the form with a model written g^m for the time rate of change of Z_p

$$\frac{dZ_1}{dt} = g(t, Z_1, Z_2), \tag{14a}$$

$$\frac{dZ_2}{dt} = g(t, Z_1, Z_2, Z_3), \tag{14b}$$

$$\vdots \tag{14c}$$

$$\frac{dZ_p}{dt} = g^m(t, Z_1, \ldots, Z_p). \tag{14d}$$

We have seen considering the Brownian motion that the choice of state-vector impacts the nature of the model and that the introduction of a stochastic term is related to the presence of a separation of scales. In some cases, the hierarchy stops, that is at a certain level p, Z_p is a function only of the first $p-1$ variables. In this case, the closure is exact and the set of equations can be considered closed. However, in the general case of nonlinear systems, the hierarchy is infinite.

We have considered so far the case of one particle described by many variables. In general, the dimension of the system (or the number of degrees of freedom), $d = \dim(\mathbf{Z})$, is given by $d = N \times p$ where N is the number of particles included in the system and p represents the number of variables attached to each particle. For this system, the complete vector which gathers all available information is then

$$\mathbf{Z} = (Z_1^1, Z_2^1, \ldots, Z_p^1; \ Z_1^2, Z_2^2, \ldots, Z_p^2; \ \ldots; \ Z_1^N, Z_2^N, \ldots, Z_p^N).$$

This vector is the state vector of the N-particle system. The vector defined by the p variables attached to each particle, $\mathbf{Z}^i = (Z_1^i, Z_2^i, \ldots, Z_p^i)$, is called the one-particle state vector, in this case for the particle labelled i. In practice the dimension of the system is huge (it might be infinite) and one has to come up with a reduced (or contracted) description, or in other words to consider a subset of dimension $d' = s \times p' \ll d$. To illustrate this problem, let us consider a N-particle system where the time evolution equation involves simply a deterministic force

$$\frac{d\mathbf{Z}(t)}{dt} = \mathbf{A}\left(t, \mathbf{Z}(t)\right).$$ (15)

The dimension of the complete state vector $\mathbf{Z}$ is equal to d, and the corresponding pdf $p(t, \mathbf{z})$ verifies the Liouville equation:

$$\frac{\partial p(t, \mathbf{z})}{\partial t} + \frac{\partial}{\partial \mathbf{z}}\left(\mathbf{A}(t, \mathbf{z})\, p(t, \mathbf{z})\right) = 0.$$ (16)

This equations is closed since in fact all the degrees of freedom of the system are explicitly tracked. We consider now a reduced pdf $p^r(t, \mathbf{z}^r)$ where $\dim(\mathbf{Z}^r) = d'$ and $p(t, \mathbf{z}) = p(t, \mathbf{z}^r, \mathbf{y})$ with, of course, $\dim(\mathbf{Y}) = d - d'$. By integration of the previous equation on $\mathbf{y}$, the transport equation for the marginal (reduced) pdf becomes

$$\frac{\partial p^r(t, \mathbf{z}^r)}{\partial t} + \frac{\partial}{\partial \mathbf{z}^r}\left[\langle \mathbf{A} \,|\, \mathbf{z}^r \rangle\, p^r(t, \mathbf{z}^r)\right] = 0,$$ (17)

where the conditional expectation is defined by,

$$\langle \mathbf{A} \,|\, \mathbf{z}^r \rangle = \int \mathbf{A}(t, \mathbf{z}^r, \mathbf{y})\, p(\mathbf{y} \,|\, t, \mathbf{z}^r)\, d\mathbf{y} = \frac{1}{p(t, \mathbf{z}^r)} \int \mathbf{A}(t, \mathbf{z}^r, \mathbf{y})\, p(t, \mathbf{z}^r, \mathbf{y})\, d\mathbf{y}.$$ (18)

Eq. (17) is now unclosed. This illustrates the fact that when a reduced description (in terms of a subset of degrees of freedom) is performed, information is lost, and one has to come up with a closure equation for higher order pdfs. In the Brownian motion exemple, we have just one particle with a different number of variable attached. In other situations, it is the number of particles which is reduced. A classical example is given by the BBGKY hierarchy (Bogoliubov, Born, Green, Kirkwood and Yvon) encountered in kinetic theory (Chapman and Cowling, 1970; Liboff, 1998).

2.4 Fast variables

In summary, we have seen that many systems are characterised by many degrees of freedom but also by the presence of several significant scales, which means that there are groups of distinct degrees of freedom characterised by very different time scales. In such a situation one says that the system has a *multiscale* character (E and Engquist, 2003) and a coarse-graining procedure can be effectively carried out in order to come up with a reduced statistical description. It is possible to say that variables characterised by very small time scales are fast, and that "slow" dynamics can be treated in terms of effective equations.

Multiscale approach is an important chapter of statistical physics with deep applications also in engineering. Castiglione et al. (2008) have recently explained in a clear a rigorous manner this approach. Here we present a brief account of the fast variable elimination which allows to start from molecular level to arrive to Brownian motion. The next section largely follows this reference (Castiglione et al., 2008), to which we refer for more details, notably the reader can find the derivation of Navier-Stokes equations from molecular level and the Smoluchowsky equation from Kramers' one.

2.5 Multiscale modelling: From molecular level to Brownian motion

We consider again the paradigmatic case of molecules. A rigorous general derivation of the Brownian motion from the first principles is a formidable task. Here we want to discuss the steps (via coarse-graining procedures) from molecular dynamics up to Brownian motion, stressing mainly the conceptual aspects (Espanol, 2004).

Consider a system of colloidal particles suspended in a liquid. At the microscopic level we introduce the canonical coordinates $(\mathbf{Q}_i, \mathbf{P}_i)$ and $(\mathbf{q}_n, \mathbf{p}_n)$ of colloidal particles and solvent molecules respectively. Omitting external potentials, the complete Hamiltonian is

$$H = \sum_i \frac{\mathbf{P}_i^2}{2M} + \sum_n \frac{\mathbf{p}_n^2}{2m} + \sum_{n,l,i,j} \left(V^{ss}(\mathbf{q}_j - \mathbf{q}_i) + V^{cc}(\mathbf{Q}_n - \mathbf{Q}_l) + V^{sc}(\mathbf{Q}_n - \mathbf{q}_i) \right),$$

(19)

where m is the mass of a solvent molecule, M is the mass of a colloidal particle (we assume $M \gg m$), V^{ss}, V^{sc} and V^{cc} are the potentials of the forces between solvent molecules, solvent and colloidal particles, colloidal particles, respectively.

The evolution of such a system is ruled by the Hamilton equations:

$$\frac{d\mathbf{Q}_i}{dt} = \frac{\partial H}{\partial \mathbf{P}_i}, \qquad \frac{d\mathbf{q}_n}{dt} = \frac{\partial H}{\partial \mathbf{p}_n},$$

(20)

$$\frac{d\mathbf{P}_i}{dt} = -\frac{\partial H}{\partial \mathbf{Q}_i}, \qquad \frac{d\mathbf{p}_n}{dt} = -\frac{\partial H}{\partial \mathbf{q}_n}.$$

The solutions of these equations give the most detailed description of the system. The problem can be simplified if one is interested on the colloidal subsystem alone. This coarsened level of description is obtained by integrating over the degrees of freedom of the solvent particles. In this case, the future state of the suspended particles is not determined by a given $\{\mathbf{Q}_i, \mathbf{P}_i\}$ configuration, but also depends on the past history of the subsystem (a unique evolution is obtained only if one knows the complete microscopic state of the system at a given time). This means that the dynamical equations for the variables $(\mathbf{Q}_i, \mathbf{P}_i)$ must contain memory effects and, in general, cannot be first order in time. However, since in comparison with the solvent molecules the colloidal particles have a much larger mass, they have a much slower evolution. Then, because of this time-scale separation between the two subsystems, and because of the huge number of the solvent particles, we can suppose that the fast solvent dynamics can be consistently decoupled from the slow colloid dynamics, by approximating its effects on the big suspended particles by means of an effective force. This force may be decomposed into a systematic part, of viscous type, and a truly stochastic fluctuating part. In such a limit of very different masses, we recover a Markovian evolution (*i.e.* first order in time) for the colloidal subsystem, that is driven by a stochastic equation:

$$
\begin{aligned}
\frac{\mathrm{d}\mathbf{Q}_i}{\mathrm{d}t} &= \mathbf{V}_i \\[2ex]
\frac{\mathrm{d}\mathbf{P}_i}{\mathrm{d}t} &= \mathbf{F}_i - \sum_j \widetilde{\zeta}_{ij} \mathbf{V}_j + \mathbf{G}_i
\end{aligned}
\tag{21}
$$

where $\mathbf{V}_i = \mathbf{P}_i/M$, $\mathbf{F}_i$ is the force on the i-th particle due to the interactions with other colloidal particles (and possibly external potentials), and $\widetilde{\zeta}_{ij}$ is the friction tensor (originating from the interaction of the solvent with the colloidal particles) that can depend on the $\mathbf{Q}$ variables. The stochastic component of the force, $\mathbf{G}_i$, is a Gaussian process with $\langle G_i^k(t) \rangle = 0$ and $\langle G_i^k(t)\, G_j^l(t') \rangle = \alpha_{ij}^{kl}\, \delta(t - t')$, where G_i^k $(k = 1, 2, 3)$ indicates the k-th spatial component of $\mathbf{G}_i$. Here $\widetilde{\alpha}_{ij}$ is a tensorial quantity in the spatial indices, like the friction tensor $\widetilde{\zeta}_{ij}$. The Fluctuation-Dissipation theorem requires that $\alpha_{ij}^{kl} = 2 k_B T \zeta_{ij}^{kl}$, where T is the temperature of the solvent, k_B the Boltzmann constant (Van Kampen, 1992).

If the colloidal suspension is dilute then we expect that the mutual influence among the colloidal particles is negligible. In this case, $\mathbf{F}_i$ is due only to possible external potentials and the solvent can be considered homogeneous, what implies that the friction tensor reduces to a scalar quantity: $\zeta_{ij}^{kl} = \zeta \delta_{ij}\delta^{kl}$, being ζ the friction coefficient. In this case eq.(21) becomes the well known Langevin equation, for the independent evolution of each colloidal particle

$$\frac{\mathrm{d}\mathbf{Q}}{\mathrm{d}t} = \mathbf{V}$$

$$\frac{\mathrm{d}\mathbf{P}}{\mathrm{d}t} = \mathbf{F}(\mathbf{Q}) - \zeta\mathbf{V} + \sqrt{2k_BT\zeta}\,\mathbf{g} \qquad (22)$$

where, for the components of the random vector $\mathbf{g}$, one has $\langle g^k(t)\rangle = 0$ and $\langle g^k(t)g^l(t')\rangle = \delta^{kl}\delta(t-t')$.

However, at this point we observe that, for typical colloidal suspension, the time scale over which the variables $\mathbf{P}$ evolve is very short, $O(10^{-6}s)$, as compared to the time scale, $O(10^{-3}s)$, of $\mathbf{Q}$ evolution. This separation of scales allows to consider another coarser level of description, only looking at the $\mathbf{Q}$ variables. In this case, a suitable equation for the colloid position variables is:

$$\frac{\mathrm{d}\mathbf{Q}_i}{\mathrm{d}t} = \frac{1}{k_BT}\sum_j \widetilde{D}_{ij}\mathbf{F}_j + \sum_j \frac{\partial}{\partial \mathbf{Q}_j}\widetilde{D}_{ij} + \mathbf{W}_i\,, \qquad (23)$$

where the force $\mathbf{F}_i$ is the same as in eq. (21), $\widetilde{D}_{ij}$ is the diffusion tensor (which, in general, depends on $\mathbf{Q}$) and $\mathbf{W}_i$ is a gaussian stochastic contribution to the velocity of the particle. We require that $\langle\mathbf{W}_i\rangle = 0$ and, by the Fluctuation-Dissipation theorem, $\langle W_i^k(t)W_j^l(t')\rangle = 2D_{ij}^{kl}\delta(t-t')$. Also in this case, for a dilute suspension equations simplify, since the diffusion tensor becomes a scalar quantity: $D_{ij}^{kl} = \delta_{ij}\delta^{kl}D$, where D is the self-diffusion coefficient of the colloidal particles. Equation (23) becomes

$$\frac{\mathrm{d}\mathbf{Q}}{\mathrm{d}t} = \frac{D}{k_BT}\mathbf{F} + \frac{\partial D}{\partial \mathbf{Q}} + \sqrt{2D}\,\mathbf{w}\,, \qquad (24)$$

where the random vector $\mathbf{w}$ enjoys the same properties as the above defined $\mathbf{g}$ vector.

In summary, the large difference between molecule and colloid masses and, the large number of molecules allow to disregard the memory effects, which can be well approximated by a stochastic markov process for the colloidal particles. Therefore, the unpredictability remains, but the future state of the subsystem only depends on its present state.

It is important to note that the coarsening procedure transforms the original deterministic problem into a stochastic one. Therefore, it is unavoidable to reason in terms of probabilities for the state of the studied (sub)system. This means that, for instance, in the first level of graining, the problem to be posed is: given the colloid system in the state $\{\mathbf{Q}_i, \mathbf{P}_i\}_0$

at time $t = 0$, what is the probability density function $p(\{\mathbf{Q}_i, \mathbf{P}_i\}, t)$ to find it in the state $\{\mathbf{Q}_i, \mathbf{P}_i\}$ at a later time t, omitting the dependence of p on the initial state[2].

So far, we have analysed the multiscale approach to this problem from a trajectory point of view. As seen in previous chapters, that has a precise correspondence with a pdf approach in the phase-space. In this framework, the probability density at time $t = 0$ is defined in the whole Γ space $p_L(t = 0) = p_L(\{\mathbf{Q}_i, \mathbf{P}_i\}_0 , \{\mathbf{q}_n, \mathbf{p}_n\}_0)$ and evolves according to the Liouville equation

$$\frac{\partial p_L}{\partial t} = -\{p_L , H\} , \tag{25}$$

where $\{p_L , H\}$ is the Poisson bracket between p_L and H, with respect to the full set of canonical variables. If one is able to find the solution $p_L(t)$, then one also has the density involving only the variables of the colloidal particles. For the first level of approximation one gets:

$$p_c(\{\mathbf{Q}_i, \mathbf{P}_i\}, t) = \int p_L(\{\mathbf{Q}_i, \mathbf{P}_i\} , \{\mathbf{q}_n, \mathbf{p}_n\}, t) \, \mathrm{d}\mathbf{q}_n \mathrm{d}\mathbf{p}_n , \tag{26}$$

and, for the second level:

$$p(\{\mathbf{Q}_i\}, t) = \int p_c(\{\mathbf{Q}_i, \mathbf{P}_i\}, t) \, \mathrm{d}\mathbf{P}_i . \tag{27}$$

Of course, when the exact solution is not attainable, one can resort to an approximate one.

In the first graining level one has to solve the evolution equation for the PdF of systems evolving according to the random dynamics (21), that is the Fokker-Planck equation (in the Ito sense, see chapter 1 (Gardiner, 1990)):

$$\frac{\partial p_c}{\partial t} = -\sum_i \left(\mathbf{V}_i \frac{\partial}{\partial \mathbf{Q}_i} + \mathbf{F}_i \frac{\partial}{\partial \mathbf{P}_i}\right) p_c + k_B T \sum_{i,j} \frac{\partial}{\partial \mathbf{P}_i} \tilde{\zeta}_{ij} \left(\frac{\partial}{\partial \mathbf{P}_j} + \frac{\mathbf{P}_j}{M k_B T}\right) p_c . \tag{28}$$

When the system (21) is a satisfying approximation of the dynamics, the solution of equation (28), with initial condition obtained by $p_L(t = 0)$, must be a good approximation of the density (26).

---

[2] Moreover, usually one does not know the initial microscopic state of a system, so that another different source of uncertainty has to be considered, in the form of a distribution on the possible initial states. This is not an uncertainty springing from the randomness of the dynamics: indeed it is present also when the detailed dynamical equations (20) drive the system. It is at the base of the ensemble point of view, à la Gibbs.

At the next level of graining, where the stochastic dynamics is given by Eq. (23), the PdF $p(\{\mathbf{Q}_i\}, t)$ of the colloidal particles positions evolves according to the so-called Smoluchowski equation:

$$\frac{\partial}{\partial t} p = -\sum_{ij} \frac{\partial}{\partial \mathbf{Q}_i} \left[\frac{\widetilde{D}_{ij} \mathbf{F}_j}{k_B T} p \right] + \sum_{ij} \frac{\partial}{\partial \mathbf{Q}_i} \widetilde{D}_{ij} \frac{\partial}{\partial \mathbf{Q}_j} p \,. \tag{29}$$

When the dilute approximation is appliable, the interesting probability densities depend only on a single particle variables, and the Fokker-Planck equations above simplify accordingly.

3   Polydispersed Turbulent Two-Phase: Formalism

3.1   Basic Principles

We have seen in the previous section that it is of capital importance to determine the relevant state vector to get an appropriate contracted description of a phenomenon. Furthermore, a multiscale approach can be followed if a separation of scales is present.

It is of course necessary to start from basic equations. We have seen that for heavy particles, equations of motion are

$$\frac{d\mathbf{U}_p}{dt} = \frac{1}{\tau_p}(\mathbf{U}_s - \mathbf{U}_p) + \mathbf{g}. \tag{30}$$

It is clear that in this case, the system is completely known if $\mathbf{x}_p, \mathbf{U}_p, \mathbf{U}_s$ are exactly given. $\mathbf{U}_s$ is the velocity fluid seen by particles and can be calculated through the solution of Navier-Stokes equations. This approach, full solution of Navier-Stokes eqs. and Lagrangian particle tracking, is in principle possible but appears out of question for many particle diameters (polydispersion), in non-trivial geometries, for high Reynolds number flows. While DNS is a formidable and useful tool to obtain physical insights, a contracted statistical description is needed. Again, we see that the starting problem is deterministic for the particle phase.

In this chapter, we want to present the modelling effort as being part of a more general framework consisting of a formalism, thus using a deductive approach rather than an inductive one, as more often encountered in literature. Thus, we introduce first a general formalism which includes both fluid and particle phase, and after we detail the state-of-the-art model form. This formalism was first put forward by Peirano and Minier (2002); Minier and Peirano (2001), and the formal approach has the merit to indicate a safe way to develop models, avoiding flaws which often characterise more heuristic approaches.

The plan of the section is the following: first, we summarise the complete formalism for both phases, fluid and particles; then we show the correspondence with average equations, as they are generally used in engineering calculations; in this framework, we put forward the state-of-the-art modelling approach. Finally, we show some numerical results.

3.2   Formalism

The first point is to define the appropriate state vector. Considering the whole system, including the fluid and the particle phase, we can give a general expression for the two-particle state vector (one fluid particle and one discrete particle). In the case of turbulent, reactive, compressible, dispersed two-phase flows, an appropriate state vector is

$$\mathbf{Z} = (\mathbf{x}_f, \mathbf{U}_f, \boldsymbol{\phi}_f, \mathbf{x}_p, \mathbf{U}_p, \boldsymbol{\phi}_p), \tag{31}$$

where $\boldsymbol{\phi}_p$ has a given dimension. We distinguish between physical space and sample space, $\mathbf{Z} = (\mathbf{y}_f, \mathbf{V}_f, \boldsymbol{\psi}_f, \mathbf{y}_p, \mathbf{V}_p, \boldsymbol{\psi}_p)$. It will be seen that $\boldsymbol{\psi}_p$ can consist of the fluid velocity seen and several scalars relevant to the discrete particles, for example diameter, enthalpy, mass fractions and so on. It is worth underlying that it is necessary to introduce two independent variables for the positions of the fluid and the discrete particles since the two kind of particles are not convected by the same velocities.

Furthermore, in any case we are considering a two-particle PDF picture (in a Lagrangian sense) of the whole system composed of the fluid and of the particles, since we are considering one particle in each phase. Generally speaking, N-particle approaches consider N particles of the same phase (for instance two fluid particles or two discrete particles, in the case of a 2-particle approach). That means considering N positions at the same time. The general state vector proposed here can be made a N-particle approach, adding in $\boldsymbol{\psi}_f, \boldsymbol{\psi}_p$ the necessary coordinates. However, while two-particle approaches may be relevant in homogeneous fluid turbulence (Pedrizzetti and Novikov, 1994), for practical purposes it is in general impossible to treat systems with more than one particle for phase. Hence, in the following, we limit ourselves to this case.

Eulerian and Lagrangian descriptions

There are two possible points of view for the description of a fluid-particle mixture. The Lagrangian one where one is interested in, at a fixed time, the probability to find two particles (a fluid particle and a discrete particle) in a given state and the Eulerian description (field approach) where one seeks the probability to find, at a given time and at two fixed points in space (a 'fluid point', $\mathbf{x}_f$, and a 'discrete-particle point', $\mathbf{x}_p$), the fluid-particle mixture in a given state.

In the case of the Lagrangian description, we define the following pdf:

$$p_{fp}^{L}(t; \mathbf{y}_f, \mathbf{V}_f, \boldsymbol{\psi}_f, \mathbf{y}_p, \mathbf{V}_p, \boldsymbol{\psi}_p), \tag{32}$$

where the probability to find a pair of particles (a fluid particle and a discrete particle) at time t, whose positions are in the range $[\mathbf{y}, \mathbf{y} + d\mathbf{y}]$, whose velocities are in the range $[\mathbf{V}, \mathbf{V} + d\mathbf{V}]$ and whose associated quantities (scalars and other variables) are in the range $[\boldsymbol{\psi}, \boldsymbol{\psi} + d\boldsymbol{\psi}]$, is

$$p_{fp}^{L}(t; \mathbf{y}_f, \mathbf{V}_f, \boldsymbol{\psi}_f, \mathbf{y}_p, \mathbf{V}_p, \boldsymbol{\psi}_p) \, d\mathbf{y}_f \, d\mathbf{V}_f \, d\boldsymbol{\psi}_f \, d\mathbf{y}_p \, d\mathbf{V}_p \, d\boldsymbol{\psi}_p. \tag{33}$$

For the field description (Eulerian point of view), the following *distribution function* is introduced:

$$p_{fp}^{E}(t, \mathbf{x}_f, \mathbf{x}_p; \mathbf{V}_f, \boldsymbol{\psi}_f, \mathbf{V}_p, \boldsymbol{\psi}_p), \tag{34}$$

where the probability to find at time t and at positions $\mathbf{x}_f$ and $\mathbf{x}_p$ the system in a given state in the range $[\mathbf{V}, \mathbf{V} + d\mathbf{V}]$ and $[\boldsymbol{\psi}, \boldsymbol{\psi} + d\boldsymbol{\psi}]$ is

$$p_{fp}^{E}(t, \mathbf{x}_f, \mathbf{x}_p; \mathbf{V}_f, \boldsymbol{\psi}_f, \mathbf{V}_p, \boldsymbol{\psi}_p) \, d\mathbf{V}_f \, d\boldsymbol{\psi}_f \, d\mathbf{V}_p \, d\boldsymbol{\psi}_p. \tag{35}$$

p_{fp}^{E} is not a PDF since, in a fluid-particle mixture, one cannot always find with probability 1, at a given time and at two different locations, a fluid and a discrete particle in any state. Therefore, there are some constraints to be applied, for a pair composed of a fluid and a discrete particle, since the two particles can not be located at the same position in physical space for a given time t. For the Lagrangian pdf when $\mathbf{y}_f = \mathbf{y}_p = \mathbf{y}$, the argument developed above implies that $p_{fp}^{L}(t; \mathbf{y}, \mathbf{V}_f, \boldsymbol{\psi}_f, \mathbf{y}, \mathbf{V}_p, \boldsymbol{\psi}_p) = 0$, and consequently, in terms of the Eulerian distribution function (for $\mathbf{x}_f = \mathbf{x}_p = \mathbf{x}$) $p_{fp}^{E}(t, \mathbf{x}, \mathbf{x}; \mathbf{V}_f, \boldsymbol{\psi}_f, \mathbf{V}_p, \boldsymbol{\psi}_p) = 0$. A direct consequence is that, at a given point $\mathbf{x}$ in physical space and a given time t, the sum of the probabilities to find a fluid particle or a discrete particle in any state is one. This can be expressed in terms of the marginals of the Eulerian distribution function as

$$\int p_{f}^{E}(t, \mathbf{x}; \mathbf{V}_f, \boldsymbol{\psi}_f) \, d\mathbf{V}_f \, d\boldsymbol{\psi}_f + \int p_{p}^{E}(t, \mathbf{x}; \mathbf{V}_p, \boldsymbol{\psi}_p) \, d\mathbf{V}_p \, d\boldsymbol{\psi}_p = 1. \tag{36}$$

where

$$p_{k}^{E}(t, \mathbf{x}_k; \mathbf{V}_k, \boldsymbol{\psi}_k) = \int p_{fp}^{E}(t, \mathbf{x}_k, \mathbf{x}_{\bar{k}}; \mathbf{V}_k, \boldsymbol{\psi}_k, \mathbf{V}_{\bar{k}}, \boldsymbol{\psi}_{\bar{k}}) \, d\mathbf{x}_{\bar{k}} \, d\mathbf{V}_{\bar{k}} \, d\boldsymbol{\psi}_{\bar{k}}, \tag{37}$$

where k is the phase index (either f or p) and $\bar{k}$ is its complement (*i.e.* for $k = p, \bar{k} = f$). Eq. (36) can also be re-written by introducing the normalization factors of p_f^E and p_p^E, namely $\alpha_f(t, \mathbf{x})$ and $\alpha_p(t, \mathbf{x})$ respectively, to yield

$$\alpha_f(t, \mathbf{x}) + \alpha_p(t, \mathbf{x}) = 1. \tag{38}$$

$\alpha_k(t, \mathbf{x})$ represents the probability to find the k phase, at time t and position $\mathbf{x}$, in any state $(0 \leq \alpha_k(t, \mathbf{x}) \leq 1)$. It is defined

$$\alpha_k(t, \mathbf{x}) = \int p_k^E(t, \mathbf{x}; \mathbf{V}_k, \boldsymbol{\psi}_k) \, d\mathbf{V}_k d\boldsymbol{\psi}_k. \tag{39}$$

As a consequence, this probability in not always one as in single-phase flows where the physical space is continuously filled by the fluid.

Marginal pdfs and mass density functions

For a complete description of the fluid particles, a mass density function $F_k^L(t; \mathbf{y}_k, \mathbf{V}_k, \boldsymbol{\psi}_k)$ is introduced where

$$F_k^L(t; \mathbf{y}_k, \mathbf{V}_k, \boldsymbol{\psi}_k) \, d\mathbf{y}_k \, d\mathbf{V}_k \, d\boldsymbol{\psi}_k, \tag{40}$$

is the probable mass of k-phase particles in an element of volume $d\mathbf{y}_k \, d\mathbf{V}_k \, d\boldsymbol{\psi}_k$.

Both mass density functions are normalized by the total mass of the respective phases, M_k

$$M_k = \int F_k^L(t; \mathbf{y}_k, \mathbf{V}_k, \boldsymbol{\psi}_k) \, d\mathbf{y}_k \, d\mathbf{V}_k \, d\boldsymbol{\psi}_k, \tag{41}$$

where the mass density functions are given by

$$F_k^L(t; \mathbf{y}_k, \mathbf{V}_k, \boldsymbol{\psi}_k) = M_k \, p_k^L(t; \mathbf{y}_k, \mathbf{V}_k, \boldsymbol{\psi}_k), \tag{42}$$

The total masses are of course defined by $M_f = \int_{\mathcal{V}_f} \rho_f(\mathbf{x}_f) d\mathbf{x}_f$ and $M_p = \sum_{i=1}^{N_p} m_{p,i}$ where N_p is the total number of discrete particles, $m_{p,i}$ the mass of the discrete particle i, and where the integration which gives M_f is performed over the domain occupied by the continuous fluid phase.

At last, we define a two-point fluid-particle mass density function

$$\mathcal{F}_{fp}^L(t; \mathbf{y}_f, \mathbf{V}_f, \boldsymbol{\psi}_f, \mathbf{y}_p, \mathbf{V}_p, \boldsymbol{\psi}_p) = M_p M_k \, p_{fp}^L(t; \mathbf{y}_f, \mathbf{V}_f, \boldsymbol{\psi}_f, \mathbf{y}_p, \mathbf{V}_p, \boldsymbol{\psi}_p), \tag{43}$$

whose marginals are related to the mass density function of the continuous phase F_f^L and the mass density function of the discrete phase F_p^L

General relations between Eulerian and Lagrangian pdfs

The quantities introduced so far are related to Lagrangian pdfs (or mdfs). Since we are interested in the derivation of field equations (Eulerian equations) for the local moments of both phases and this can only be done by means of Eulerian tools, we need to provide the link between the Lagrangian quantities introduced and the Eulerian ones.

Following Balescu (1997), we can write

$$
\begin{aligned}
\mathcal{F}^E_{fp}&(t,\mathbf{x}_f,\mathbf{x}_p;\mathbf{V}_f,\boldsymbol{\psi}_f,\mathbf{V}_p,\boldsymbol{\psi}_p) \\
&= \mathcal{F}^L_{fp}(t;\mathbf{y}_f = \mathbf{x}_f, \mathbf{V}_f, \boldsymbol{\psi}_f, \mathbf{y}_p = \mathbf{x}_p, \mathbf{V}_p, \boldsymbol{\psi}_p) \\
&= \int \mathcal{F}^L_{fp}(t;\mathbf{y}_f, \mathbf{V}_f, \boldsymbol{\psi}_f, \mathbf{y}_p, \mathbf{V}_p, \boldsymbol{\psi}_p)\, \delta(\mathbf{x}_f - \mathbf{y}_f)\, \delta(\mathbf{x}_p - \mathbf{y}_p)\, d\mathbf{y}_f\, d\mathbf{y}_p,
\end{aligned}
\tag{44}
$$

where $\mathcal{F}^E_{fp}$ is the Eulerian two-point fluid-particle mass density function. By direct integration of the previous equation, relations for the associated marginals, the Eulerian one-point k-phase mass density function, $\mathcal{F}^E_k$ are obtained

$$
\begin{aligned}
\mathcal{F}^E_k(t,\mathbf{x}_k;\mathbf{V}_k,\boldsymbol{\psi}_k) &= \mathcal{F}^L_k(t;\mathbf{y}_k = \mathbf{x}_k, \mathbf{V}_k, \boldsymbol{\psi}_k) \\
&= \int \mathcal{F}^L_k(t;\mathbf{y}_k, \mathbf{V}_k, \boldsymbol{\psi}_k)\, \delta(\mathbf{x}_k - \mathbf{y}_k)\, d\mathbf{y}_k,
\end{aligned}
\tag{45}
$$

By recalling that $\mathcal{F}^L_k = M_{\bar{k}} F^L_k$, the relation implies $\mathcal{F}^E_k = M_{\bar{k}} F^E_k$. Therefore the relations between Eulerian and Lagrangian mass density functions F_k are also given by eq. (45).

The basic elements are now given. In order to derive mean equations there are two equivalents strategies: starting always from Lagrangian two-point pdf p^L_{fp}, in the first, relations between Eulerian and Lagrangian mdfs are worked out using,

$$
\begin{aligned}
&\mathcal{F}^E_{fp}(t;\mathbf{x}_f = \mathbf{x}, \mathbf{x}_p;\mathbf{V}_f,\boldsymbol{\psi}_f,\mathbf{V}_p,\boldsymbol{\psi}_p), \\
&\mathcal{F}^E_{fp}(t;\mathbf{x}_f, \mathbf{x}_p = \mathbf{x};\mathbf{V}_f,\boldsymbol{\psi}_f,\mathbf{V}_p,\boldsymbol{\psi}_p).
\end{aligned}
\tag{46}
$$

These two mdfs can give both marginals at the same point in physical space, i.e. $F^E_k(t,\mathbf{x};\mathbf{V}_k,\boldsymbol{\psi}_k)$.

In the second, relations between Eulerian and Lagrangian mdfs are worked out at the one-point pdf level, with Eq. (42) and Eqs. (45). Information is obtained in the form of the one-point k-phase mass density functions, $F^E_k(t,\mathbf{x};\mathbf{V}_k,\boldsymbol{\psi}_k)$, from p^L_k.

Two-point relations between Eulerian and Lagrangian pdfs

Following the first procedure, one can write

$$\mathcal{F}_{fp}^{E}(t,\mathbf{x}_f,\mathbf{x}_p;\mathbf{V}_f,\boldsymbol{\psi}_f,\mathbf{V}_p,\boldsymbol{\psi}_p) =$$
$$\int p|_{fp}^{L}(t;\mathbf{x}_f,\mathbf{V}_f,\boldsymbol{\psi}_f,\mathbf{x}_p,\mathbf{V}_p,\boldsymbol{\psi}_p|\, t_0;\mathbf{x}_{f0},\mathbf{V}_{f0},\boldsymbol{\psi}_{f0},\mathbf{x}_{p0},\mathbf{V}_{p0},\boldsymbol{\psi}_{p0})$$
$$\mathcal{F}_{fp}^{E}(t,\mathbf{x}_{f0},\mathbf{x}_{p0};\mathbf{V}_{f0},\boldsymbol{\psi}_{f0},\mathbf{V}_{p0},\boldsymbol{\psi}_{p0})\,d\mathbf{x}_{f0}\,d\mathbf{V}_{f0}\,d\boldsymbol{\psi}_{f0}\,d\mathbf{x}_{p0}\,d\mathbf{V}_{p0}\,d\boldsymbol{\psi}_{p0}.$$
$$(47)$$

This relation shows that the Eulerian mass density function $\mathcal{F}_{fp}^{E}$ is 'propagated' by the transitional pdf, that is the transitional pdf $p|_{fp}^{L}$ is the *propagator* of the two-point fluid-particle Eulerian mass density function. Consequently the partial differential equation which is verified by the transitional pdf is also verified by the Eulerian mass density function $\mathcal{F}_{fp}^{E}$.

The definitions of the expected densities, $\langle\rho_f\rangle(t,\mathbf{x})$ and $\langle\rho_p\rangle(t,\mathbf{x})$, and the probability of presence of both phases $\alpha_f(t,\mathbf{x})$ and $\alpha_p(t,\mathbf{x})$, can be expressed in terms of the two-point Eulerian mdf.

$$\alpha_k(t,\mathbf{x})\langle\rho_k\rangle(t,\mathbf{x}) =$$
$$\frac{1}{M_{\bar{k}}}\int \mathcal{F}_{fp}^{E}(t,\mathbf{x},\mathbf{x}_p;\mathbf{V}_f,\boldsymbol{\psi}_f,\mathbf{V}_p,\boldsymbol{\psi}_p)\,d\mathbf{x}_{\bar{k}}\,d\mathbf{V}_p\,d\boldsymbol{\psi}_p\,d\mathbf{V}_f\,d\boldsymbol{\psi}_f,$$
$$\alpha_k(t,\mathbf{x}) =$$
$$\frac{1}{M_{\bar{k}}}\int \frac{1}{\rho_f(\boldsymbol{\psi}_f)}\,\mathcal{F}_{fp}^{E}(t,\mathbf{x},\mathbf{x}_p;\mathbf{V}_f,\boldsymbol{\psi}_f,\mathbf{V}_p,\boldsymbol{\psi}_p)\,d\mathbf{x}_{\bar{k}}\,d\mathbf{V}_p\,d\boldsymbol{\psi}_p\,d\mathbf{V}_f\,d\boldsymbol{\psi}_f$$
$$(48)$$

Finally, the expected densities and the probability of presence of both phases can be written in terms of the marginals of $\mathcal{F}_{fp}^{E}$

$$\alpha_k(t,\mathbf{x})\langle\rho_k\rangle(t,\mathbf{x}) = \int F_k^{E}(t,\mathbf{x};\mathbf{V}_k,\boldsymbol{\psi}_k)\,d\mathbf{V}_k\,d\boldsymbol{\psi}_k,$$
$$\alpha_k(t,\mathbf{x}) = \int \frac{1}{\rho_k(\boldsymbol{\psi}_k)}\,F_k^{E}(t,\mathbf{x};\mathbf{V}_k,\boldsymbol{\psi}_k)\,d\mathbf{V}_k\,d\boldsymbol{\psi}_k,$$
$$(49)$$

Relations between Eulerian and Lagrangian marginals

Following the second procedure, Using Eq. (45), the definition of the Lagrangian mdf $F_k^{L} = M_k\,p_k^{L}$, and introducing the fluid transitional pdf $p|_f^{L}$, one can write

$$F_k^E(t, \mathbf{x}; \mathbf{V}_k, \boldsymbol{\psi}_k) = \int p|_k^L(t; \mathbf{x}, \mathbf{V}_k, \boldsymbol{\psi}_k | t_0; \mathbf{x}_{k0}, \mathbf{V}_{k0}, \boldsymbol{\psi}_{k0})$$
$$F_k^E(t, \mathbf{x}_0; \mathbf{V}_{k0}, \boldsymbol{\psi}_{k0}) \, d\mathbf{x}_0 \, d\mathbf{V}_{k0} \, d\boldsymbol{\psi}_{k0}. \tag{50}$$

This relation shows that the fluid Eulerian mass density function F_k^E is 'propagated' by the fluid transitional pdf.

The integral of F_k^E over phase space $(\mathbf{V}_k, \boldsymbol{\psi}_k)$ is the expected density at $(t, \mathbf{x})$, denoted $\langle \rho_k \rangle(t, \mathbf{x})$, which is defined by the following equation

$$\alpha_k(t, \mathbf{x}) \langle \rho_k \rangle(t, \mathbf{x}) = \int \rho_k(\psi_k) \, p_k^E(t, \mathbf{x}; \mathbf{V}_k, \boldsymbol{\psi}_k) d\mathbf{V}_k d\boldsymbol{\psi}_k, \tag{51}$$

where the Eulerian mass density function F_k^E is given by

$$F_k^E(t, \mathbf{x}; \mathbf{V}_k, \boldsymbol{\psi}_k) = \rho(\psi_k) \, p_k^E(t, \mathbf{x}; \mathbf{V}_k, \boldsymbol{\psi}_k). \tag{52}$$

Integration of F_k^L and F_k^E over phase space $(\mathbf{V}_k, \boldsymbol{\psi}_k)$ yields (where the notation $\mathbf{y} = \mathbf{x}$ in the Lagrangian pdfs is, from now on, dropped most of the time for the sake of clarity)

$$p_k^L(t; \mathbf{x}) = \frac{1}{M_k} \, \alpha_k(t, \mathbf{x}) \langle \rho_k \rangle(t, \mathbf{x}), \tag{53}$$

and therefore the conditional expectation $p_k^L(t; \mathbf{V}_k, \boldsymbol{\psi}_k | \mathbf{x})$ is given by

$$p_k^L(t; \mathbf{V}_k, \boldsymbol{\psi}_k | \mathbf{x}) = \frac{\rho_k(\psi_k)}{\alpha_k(t, \mathbf{x}) \langle \rho_k \rangle(t, \mathbf{x})} \, p^E(t, \mathbf{x}; \mathbf{V}_k, \boldsymbol{\psi}_k). \tag{54}$$

Therefore, we find that in a compressible flow, the fluid Lagrangian pdf conditioned by the position is not the fluid Eulerian distribution function but the density-weighted fluid Eulerian pdf, p_k^E / α_k. Furthermore, even in an incompressible flow, particles are in general a compressible phase.

PDF equation in dispersed two-phase flows

The existence of a *propagator* ($p|_{fp}^L$ for the fluid-particle mixture or $p|_k^L$ for the phase k) indicates that the Lagrangian point of view is the natural choice and from now on this point of view is retained. No emphasis is put on modeling for the moment. This will be discussed in the following section.

For the sake of simplicity, only dispersed two-phase flows with two-way coupling are considered. The possible influence of collisional mechanisms (between discrete particles) can be accounted for in the frame of the present formalism, but it is not discussed here. The reader can find some details elsewhere (Minier and Peirano, 2001). In order to write the partial differential equation verified by the *propagator*, we recall the exact equations for the trajectories of fluid and discrete particles:

$$
\begin{aligned}
dx_{f,i}^{+} &= U_{f,i}^{+}\, dt, \\
dU_{f,i}^{+} &= A_{f,i}^{+}\, dt + A_{p\to f}^{+}(t,\mathbf{Z},\langle\mathbf{Z}\rangle)\, dt, \\
d\phi_{f,l}^{+} &= \Gamma_f \Delta\phi_{f,l}^{+}\, dt + S_f(\boldsymbol{\phi_f^{+}})\, dt, \\
dx_{p,i}^{+} &= U_{p,i}^{+}\, dt, \\
dU_{p,i}^{+} &= A_{p,i}^{+}\, dt, \\
d\phi_{p,k}^{+} &= \Gamma_p \Delta\phi_{p,k}^{+}\, dt + S_p(\boldsymbol{\phi_p^{+}})\, dt,
\end{aligned}
\tag{55}
$$

where the indexes l and k refer to the dimensions of $\boldsymbol{\phi_f}$ and $\boldsymbol{\phi_p}$, respectively. As in the previous sections, the $+$ subscript is used to indicate the exact trajectories in contrast to the modelled ones. Exact accelerations are given by:

$$
\begin{aligned}
A_{f,i}^{+} &= -\frac{1}{\rho_f}\frac{\partial P^{+}}{\partial x_i} + \nu\Delta U_{f,i}^{+}, \\
A_{p,i}^{+} &= \frac{1}{\tau_p}(U_{s,i}^{+} - U_{p,i}^{+}) + g_i.
\end{aligned}
\tag{56}
$$

A new term is added in the momentum equation of the fluid to account for the influence of the particles on the fluid. The exact expression for this acceleration, which is induced by the presence of the discrete particles, is not a priori known and possible models for the trajectories of stochastic fluid particles are discussed by Minier and Peirano (2001) and by Peirano et al. (2006).

Using the techniques presented in Chapter 1, the transitional pdf $p|_{fp}^{L+}$ verifies the following partial differential equation

$$
\begin{aligned}
\frac{\partial p|_{fp}^{L+}}{\partial t} &+ V_{f,i}\frac{\partial p|_{fp}^{L+}}{\partial y_{f,i}} + V_{p,i}\frac{\partial p|_{fp}^{L+}}{\partial y_{p,i}} = -\frac{\partial}{\partial V_{f,i}}(\langle A_{f,i}^{+}|\mathbf{Z}=\mathbf{z}\rangle\, p|_{fp}^{L+}) \\
&-\frac{\partial}{\partial V_{f,i}}(\langle A_{p\to f}^{+}(t,\mathbf{Z},\langle\mathbf{Z}\rangle)|\mathbf{Z}=\mathbf{z}\rangle\, p|_{fp}^{L+}) - \frac{\partial}{\partial V_{p,i}}(\langle A_{p,i}^{+}|\mathbf{Z}=\mathbf{z}\rangle\, p|_{fp}^{L+}) \\
&-\frac{\partial}{\partial \psi_{f,l}}(\langle \Gamma_f \Delta\phi_{f,l}^{+}|\mathbf{Z}=\mathbf{z}\rangle\, p|_{fp}^{L+}) - \frac{\partial}{\partial \psi_{f,l}}(S_{f,l}(\boldsymbol{\psi_f}^{+})\, p|_{fp}^{L+}) \\
&-\frac{\partial}{\partial \psi_{p,k}}(\langle \Gamma_p \Delta\phi_{p,k}^{+}|\mathbf{Z}=\mathbf{z}\rangle\, p|_{fp}^{L+}) - \frac{\partial}{\partial \psi_{p,k}}(S_{p,k}(\boldsymbol{\psi_p}^{+})\, p|_{fp}^{L+}),
\end{aligned}
\tag{57}
$$

Then, introducing the definition of the exact fluid-particle pdf in terms of the transitional pdf

$$p_{fp}^{L+}(t;\mathbf{x}_f,\mathbf{V}_f,\boldsymbol{\psi}_f,\mathbf{x}_p,\mathbf{V}_p,\boldsymbol{\psi}_p) =$$

$$\int p|_{fp}^{L+}(t;\mathbf{x}_f,\mathbf{V}_f,\boldsymbol{\psi}_f,\mathbf{x}_p,\mathbf{V}_p,\boldsymbol{\psi}_p|\,t_0;\mathbf{x}_{f0},\mathbf{V}_{f0},\boldsymbol{\psi}_{f0},\mathbf{x}_{p0},\mathbf{V}_{p0},\boldsymbol{\psi}_{p0})$$

$$\tag{58}$$

$$p_{fp}^{L+}(t;\mathbf{x}_{f0},\mathbf{V}_{f0},\boldsymbol{\psi}_{f0},\mathbf{x}_{p0},\mathbf{V}_{p0},\boldsymbol{\psi}_{p0})\,d\mathbf{x}_{f0}\,d\mathbf{V}_{f0}\,d\boldsymbol{\psi}_{f0}\,d\mathbf{x}_{p0}\,d\mathbf{V}_{p0}\,d\boldsymbol{\psi}_{p0},$$

and using Eq. (47), it can be shown that Eq. (57) is satisfied by the the Lagrangian pdf, p_{fp}^{L+}, and by the Eulerian mass density function, $\mathcal{F}_{fp}^{E+}$. From this equation, it is therefore possible to derive all mean equations.

General modelling guidelines

Even if the details of the modeling are postponed, it essential here to anticipate at least the form of the model, before deriving mean equations. Eq. (57) is exact, and if all the relevant variables were included in the state vector, it could be considered closed. As explained before, this would involve a N-particles approach which is not the scope of present discussion. A much coarser state vector is to be considered.

As we will see in more details soon, Kolmogorov theory tells us that the acceleration of fluid particles (Monin and Yaglom, 1975) and the acceleration of the fluid sampled along discrete particle trajectories are fast variables (with dt being the reference time scale). Both accelerations are external variables which have to be modelled and the two-particle (one fluid particle and one discrete particle) state vector is defined by

$$\mathbf{Z} = (\mathbf{x}_f,\mathbf{U}_f,\mathbf{x}_p,\mathbf{U}_p,\mathbf{U}_s,d_p), \tag{59}$$

and in sample space, $(\mathbf{x},\mathbf{U},d_p) \leftrightarrow (\mathbf{y},\mathbf{V},\delta_p)$. We have neglected the reactive variables, since the extension is straightforward. The selection of the velocity of the fluid seen by the discrete particles, namely $\mathbf{U}_s$, as an independent variable linked to the discrete particles is a noteworthy point. One could thus wonder whether the introduction of the velocity of the fluid seen as an independent variable is justified when $\mathbf{U}_f$ is already included. Yet, the statistics of the velocity of the fluid seen are different from the statistics of a fluid particle due to particle inertia and crossing-trajectory effects. Furthermore, these two fluid velocities do not correspond to the same trajectories: $\mathbf{U}_f$ is the velocity along the fluid particle trajectory $\mathbf{x}_f$, while $\mathbf{U}_s$ is the velocity along the discrete particle trajectory $\mathbf{x}_p$. That justifies the presence of these two fluid velocities which are treated as independent variables.

Stochastic model for $\mathbf{U}_f$ and $\mathbf{U}_s$

In the trajectory point of view, an useful approach is to use a Langevin equation. Then, the Langevin equation model consists in writing the increments in time of $\mathbf{U}_f$ and $\mathbf{U}_s$ as diffusion processes,

$$dx_{f,i} = U_{f,i}\, dt, \tag{60a}$$

$$dU_{f,i} = [A_{f,i}(t,\mathbf{Z}) + A_{p\to f,i}(t,\mathbf{Z},\langle\mathbf{Z}\rangle)]\, dt + B_{f,ij}(t,\mathbf{Z})\, dW'_j, \tag{60b}$$

$$dx_{p,i} = U_{p,i}\, dt, \tag{60c}$$

$$dU_{p,i} = A_{p,i}(t,\mathbf{Z}),\, dt \tag{60d}$$

$$dU_{s,i} = [A_{s,i}(t,\mathbf{Z}) + A_{p\to s,i}(t,\mathbf{Z},\langle\mathbf{Z}\rangle)]\, dt + B_{s,ij}(t,\mathbf{Z})\, dW_j, \tag{60e}$$

where the drift vectors $A_{f,i}, A_{p\to f,i}$, $A_{p\to s,i}$, $A_{s,i}$ and the diffusion matrices $B_{f,ij}$, $B_{s,ij}$ have to be modelled. Hereafter we consider $d_p = $ constant. The details concerning drift and diffusion terms are not necessary at this level. It is worth noting just that the two-way coupling term enters the equations of $\mathbf{U}_f$ and $\mathbf{U}_s$, $\mathbf{A}_{p\to f}$ which reflects the influence of the discrete particles on the fluid. This is a simple consequence of Newton's third law: the fluid exerts a force $\mathbf{F}_{f\to p}$ on the discrete particles and, in return, the particles exert a force $\mathbf{F}_{p\to f} = -\mathbf{F}_{f\to p}$ on the fluid. The force exerted by one particle on the fluid corresponds to the drag force written here as

$$\mathbf{F}_{p\to f} = -m_p\mathbf{A}_p^D = -m_p\frac{\mathbf{U}_s - \mathbf{U}_p}{\tau_p}. \tag{61}$$

The issue of two-way coupling is out of the present scope and the reader can find details in Minier and Peirano (2001) and in Peirano et al. (2006).

Summarising, the complete Langevin equation model is equivalent to a Fokker-Planck equation given in closed form for the transitional pdf, $p|_{fp}^{L}$. As demonstrated previously, the Fokker-Planck equation verified by $p|_{fp}^{L}$ is also verified by the two-point fluid-particle Eulerian mass density function $\mathcal{F}_{fp}^{E}$ and the two-point fluid-particle Lagrangian pdf p_{fp}^{L}. This Fokker-Planck equation is, for the transitional pdf

$$\frac{\partial p|_{fp}^{L}}{\partial t} + V_{f,i}\frac{\partial p|_{fp}^{L}}{\partial y_{f,i}} + V_{p,i}\frac{\partial p|_{fp}^{L}}{\partial y_{p,i}} = -\frac{\partial}{\partial V_{f,i}}\left([A_{f,i} + \langle A_{p\to f,i}\,|\,\mathbf{y}_f,\mathbf{V}_f\rangle]\, p|_{fp}^{L}\right)$$

$$-\frac{\partial}{\partial V_{p,i}}\left(A_{p,i}\, p|_{fp}^{L}\right) - \frac{\partial}{\partial V_{s,i}}\left([A_{s,i} + \langle A_{p\to s,i}\,|\,\mathbf{y}_p,\mathbf{V}_p,\boldsymbol{\psi}_p\rangle]\, p|_{fp}^{L}\right)$$

$$+\frac{1}{2}\left(\frac{\partial^2}{\partial V_{f,i}\partial V_{f,j}}([B_f B_f^T]_{ij}\, p|_{fp}^{L}) + \frac{1}{2}\frac{\partial^2}{\partial V_{s,i}\partial V_{s,j}}([B_s B_s^T]_{ij}\, p|_{fp}^{L}\right).$$

$$\tag{62}$$

Computations of the two-point fluid-particle pdf p_{fp}^L can now be performed using the trajectory point of view, that is by Lagrangian/Lagrangian simulations. Time evolution equations, Eqs. (60) are written for an ensemble made up of fluid and discrete particles which are tracked together. Both have specified variables attached to them which appear as independent variables in the pdf.

By direct integration of Eq. (62) the Fokker-Planck equations verified by the one-particle transitional pdfs (the fluid transitional pdf $p|_f^L$ and the particle transitional pdf $p|_p^L$) can be obtained. In the case of the fluid transitional pdf, $p|_f^L$,

$$
\frac{\partial p|_f^L}{\partial t} + V_{f,i}\frac{\partial p|_f^L}{\partial y_{f,i}} = -\frac{\partial}{\partial V_{f,i}}(A_{f,i}\,p|_f^L)
$$
$$
- \frac{\partial}{\partial V_{f,i}}(\langle A_{p\to f,i}\,|\,\mathbf{y}_f,\mathbf{V}_f\rangle\,p|_f^L) + \frac{1}{2}\frac{\partial^2}{\partial V_{f,i}\partial V_{f,j}}([B_f B_f^T]_{ij}\,p|_f^L).
$$
$$(63)$$

When two-way coupling is accounted for, the two-point (one fluid particle and one discrete particle) information is necessary since the dynamics of the fluid phase involve variables attached to the discrete particles. In the case of the particle transitional pdf, $p|_p^L$,

$$
\frac{\partial p|_p^L}{\partial t} + V_{p,i}\frac{\partial p|_p^L}{\partial y_{p,i}} = -\frac{\partial}{\partial V_{p,i}}(A_{p,i}\,p|_p^L)
$$
$$
- \frac{\partial}{\partial V_{s,i}}\big([A_{s,i} + \langle A_{p\to s,i}\,|\,\mathbf{y}_p,\mathbf{V}_p,\boldsymbol{\psi}_p\rangle]\,p|_p^L\big) + \frac{1}{2}\frac{\partial^2}{\partial V_{p,i}\partial V_{p,j}}([B_p B_p^T]_{ij}\,p|_p^L),
$$
$$(64)$$

where $\boldsymbol{\psi}_p \leftrightarrow (\mathbf{V}_s,\boldsymbol{\delta}_p)$.

Mean field equations

The use of the two-point fluid-particle pdf allows an equal treatment of both phases and it is a compact way to derive a set of field equations. These field equations are often referred to as the 'Eulerian model' or sometimes 'two-fluid model'. Here, we would like to call it a *two-field model*: this term describes the spirit of the approach which is to derive field equations for both phases. Now, let us discuss how such equations are derived. We consider here the particular case of one-way coupling, *i.e.* $\mathbf{A}_{p\to f} = \mathbf{A}_{p\to s} = 0$ (for incompressible flows with particles of constant density but variable diameter), for the sake of simplicity. The case of two-way coupling does not present any difficulty from a formal point of view, and it has been addressed (Minier and Peirano, 2001; Peirano and Minier, 2002; Peirano et al., 2006).

Let us recall that two possible strategies are possible to derive mean equation: from the two-point mass density function or from the respective one-point FDF. In the first, it is indeed possible to keep the joint (one fluid point-one particle point) information for the field description by treating the two-point fluid-particle Eulerian mass density function, $\mathcal{F}_{fp}^{E}$. We study, as mentioned previously, the cases where $\mathbf{x}_f = \mathbf{x}$ for the fluid, and $\mathbf{x}_p = \mathbf{x}$ for the discrete phase, that is the following mass density functions

$$\begin{aligned}
\mathcal{F}_{fp}^{E}(t, \mathbf{x}_f = \mathbf{x}, \mathbf{x}_p; \mathbf{V}_f, \boldsymbol{\psi}_f, \mathbf{V}_p, \boldsymbol{\psi}_p), \\
\mathcal{F}_{fp}^{E}(t, \mathbf{x}_f, \mathbf{x}_p = \mathbf{x}; \mathbf{V}_f, \boldsymbol{\psi}_f, \mathbf{V}_p, \boldsymbol{\psi}_p).
\end{aligned} \tag{65}$$

By direct integration, the Fokker-Planck equations verified by the marginals $\mathcal{F}_{f}^{E}$ and $\mathcal{F}_{p}^{E}$ can be obtained from the Fokker-Planck equation verified by $\mathcal{F}_{pf}^{E}$ which is, in its turn, obtained from the partial differential equation verified by the transitional pdf $p|_{fp}^{L}$. The latter equations are also verified by F_{f}^{E} and F_{p}^{E}.

In the second, F_{f}^{L} and F_{p}^{L} have been obtained as well as their correspondence with the field (Eulerian) description could be made. After that, we have found that each Eulerian mass density function, F_{f}^{E} and F_{p}^{E}, is propagated by the corresponding transitional pdf ($p|_{f}^{L}$ and $p|_{p}^{L}$ respectively). This allows us to write immediately the Fokker-Planck (partial differential) equation verified by F_{f}^{E} and F_{p}^{E} from the Fokker-Planck equations verified by the transitional pdfs $p|_{f}^{L}$ and $p|_{p}^{L}$ or from the Fokker-Planck equation verified by the transitional pdf $p|_{fp}^{L}$. From that, averaging mean field equations can be written.

Fluid and discrete particle expectations

In the present case (discrete particle of constant density but variable diameter in an incompressible flow) all information is contained in the distribution functions $p_{p}^{E}(t, \mathbf{x}; \mathbf{V}_p, \boldsymbol{\psi}_p)$ (with $\boldsymbol{\psi}_p = (\mathbf{V}_s, \delta_p)$ for the discrete phase and $p_{f}^{E}(t, \mathbf{x}; \mathbf{V}_f)$ for the fluid but the derivation will be addressed in terms of the mass density function $F_{k}^{E}(t, \mathbf{x}; \mathbf{V}_k, \boldsymbol{\psi}_k) = \rho_k \, p_{k}^{E}(t, \mathbf{x}; \mathbf{V}_k, \boldsymbol{\psi}_k)$ for both phases. The mathematical definition of the expected Eulerian value of a function $H(\mathbf{V}_k, \boldsymbol{\psi}_k)$ (a sufficiently smooth function attached to a given particle, i.e., a fluid or a discrete particle) is

$$\alpha_k(t, \mathbf{x}) \langle \rho_k \rangle \langle H \rangle (t, \mathbf{x}) = \int H \, F_{k}^{E}(t, \mathbf{x}; \mathbf{V}_k, \boldsymbol{\psi}_k) \, d\mathbf{V}_k \, d\boldsymbol{\psi}_k, \tag{66}$$

Therefore, in the present formalism, all expected values must be understood as mass-weighted mean values. All the velocity moments can be obtained accordingly. For instance, the fluid seen-particle velocity moments (of order $n + m$) can be defined

$$
\alpha_p(t, \mathbf{x})\rho_p\langle u_{s,i_1} \ldots u_{s,i_n} u_{p,j_1} \ldots u_{p,j_m}\rangle(t, \mathbf{x}) = \int \prod_{k=1}^{n} v_{s,i_k} \prod_{l=1}^{m} v_{p,j_l}\, F_p^E(t, \mathbf{x}; \mathbf{V}_p, \boldsymbol{\psi}_p)\, d\mathbf{V}_p\, d\boldsymbol{\psi}_p, \quad (67)
$$

where $j_l \in \{1, 2, 3\}\ \forall l$. Considering also the diameter in the state-vector, a general definition is introduced, that is a moment of order $n + m + q$,

$$
\alpha_p(t, \mathbf{x})\rho_p\langle (d_p')^n\, u_{s,i_1} \ldots u_{s,i_m} u_{p,j_1} \ldots u_{p,j_q}\rangle(t, \mathbf{x}) = \int (\delta_p')^n \prod_{k=1}^{m} v_{s,i_k} \prod_{l=1}^{q} v_{p,j_l}\, F_p^E(t, \mathbf{x}; \mathbf{V}_p, \boldsymbol{\psi}_p)\, d\mathbf{V}_p\, d\boldsymbol{\psi}_p, \quad (68)
$$

Different moments can be easily computed. For instance, the velocity moments of order n read

$$
\alpha_f(t, \mathbf{x})\langle u_{f,i_1} \ldots u_{f,i_n}\rangle(t, \mathbf{x}) = \int \prod_{k=1}^{n} v_{f,i_k}\, p_f^E(t, \mathbf{x}; \mathbf{V}_f)\, d\mathbf{V}_f. \quad (69)
$$

It is now necessary to clarify the correspondence between the mathematical expectations, Eq.(66), and Monte Carlo estimations drawn from a finite ensemble of particles. It is easy to show (Minier and Peirano, 2001; Peirano and Minier, 2002) that

$$
\alpha_p(t, \mathbf{x})\, \rho_p\langle H_p\rangle \simeq \frac{1}{\delta\mathcal{V}_x} \sum_{i=1}^{N_x^p} m_p^i H(\mathbf{U}_p^i(t), \boldsymbol{\phi}_p^i(t)), \quad (70)
$$

and with

$$
\alpha_p(t, \mathbf{x})\, \rho_p \simeq \frac{\sum_{i=1}^{N_x^p} m_p^i}{\delta\mathcal{V}_x}, \quad (71)
$$

we have

$$
\langle H_p\rangle \simeq H_{p,N} = \frac{\sum_{i=1}^{N_x^p} m_p^i H(\mathbf{U}_p^i(t), \boldsymbol{\phi}_p^i(t))}{\sum_{i=1}^{N_x^p} m_p^i}. \quad (72)
$$

For a polydispersed particle phase, particles have different masses even when ρ_p is constant since their diameter varies. The important consequence is that, even for constant density particles, the natural definition or understanding of a mean quantity is the *mass-weighted*-average, or the volume-weighted average when ρ_p is constant.

Field equations for the fluid phase

With Eqs. (50) and (63), it is straightforward to write the Fokker-Planck equation verified by $F_f^E(t, \mathbf{x}; \mathbf{V}_f)$

$$\frac{\partial F_f^E}{\partial t} + V_{f,i} \frac{\partial F_f^E}{\partial x_{f,i}} = -\frac{\partial}{\partial V_{f,i}}(A_{f,i}\, F_f^E) + \frac{1}{2} \frac{\partial^2}{\partial V_{f,i}\partial V_{f,j}}([B_f B_f^T]_{ij}\, F_f^E). \tag{73}$$

In the case of constant fluid density, ρ_f, the same equation is verified by $p_f^E(t, \mathbf{x}; \mathbf{V}_f)$. It is more convenient to make a change of coordinates in velocity space, $\mathbf{v}_f = \mathbf{V}_f - \langle \mathbf{U}_f\rangle(t, \mathbf{x})$, and it is straightforward to prove that the Fokker-Planck equation verified by $p_f^E(t, \mathbf{x}; \mathbf{v}_f)$ reads

$$\begin{aligned}
\frac{dp_f^E}{dt} + v_{f,i}\frac{\partial p_f^E}{\partial x_i} = &-\frac{\partial}{\partial v_{f,i}}(A_{f,i}\, p_f^E) + \frac{1}{2}\frac{\partial^2}{\partial v_{f,i} v_{f,j}}\left((B_f B_f^T)_{ij}\, p_f^E\right) \\
&+ \frac{d\langle U_{f,i}\rangle}{dt}\frac{\partial p_f^E}{\partial v_{f,i}} + v_{f,i}\frac{\partial \langle U_{f,j}\rangle}{\partial x_i}\frac{\partial p_f^E}{\partial v_{f,j}},
\end{aligned} \tag{74}$$

where $d/dt = \partial/\partial t + \langle U_{f,i}\rangle \partial/\partial x_i$ is the Eulerian derivative along the path of a fluid particle.

Let us multiply Eq. (74) by $\mathcal{H}_f$ and apply the $\langle \cdot \rangle$ operator, Eq. (66). Assuming that all generalized integrals converge to zero in the limit $v_{f,i} \to \pm\infty$ and that uniform convergence is verified, after some derivations, one finds

$$\begin{aligned}
\frac{d}{dt}(\alpha_f \langle \mathcal{H}_f\rangle) + &\frac{\partial}{\partial x_i}(\alpha_f \langle v_{f,i}\, \mathcal{H}_f\rangle) = \alpha_f \langle A_{f,i}\frac{\partial \mathcal{H}_f}{\partial v_{f,i}}\rangle + \frac{1}{2}\alpha_f \langle (B_f B_f^T)_{ij}\frac{\partial^2 \mathcal{H}_f}{\partial v_{f,i}\partial v_{f,j}}\rangle \\
&- \alpha_f \frac{d\langle U_{f,i}\rangle}{dt}\langle \frac{\partial \mathcal{H}_f}{\partial v_{f,i}}\rangle - \alpha_f \frac{\partial \langle U_{f,j}\rangle}{\partial x_i}\langle \frac{\partial (v_{f,i}\mathcal{H}_f)}{\partial v_{f,j}}\rangle.
\end{aligned} \tag{75}$$

By replacing $\mathcal{H}_f$ by $\mathcal{H}_f = 1$, $\mathcal{H}_f = V_{f,i}$ and $\mathcal{H}_f = v_{f,i}v_{f,j}$, the continuity equation, the momentum equations and the Reynolds-stress equations are obtained, respectively. Multiplying these equations by ρ_f yields for the continuity equation,

$$\frac{\partial}{\partial t}(\alpha_f \rho_f) + \frac{\partial}{\partial x_i}(\alpha_f\, \rho_f \langle U_{f,i}\rangle) = 0, \tag{76}$$

for the momentum equation

$$\alpha_f \rho_f \frac{d}{dt}\langle U_{f,i}\rangle = -\frac{\partial}{\partial x_j}(\alpha_f\, \rho_f \langle u_{f,i}u_{f,j}\rangle) + \alpha_f\, \rho_f \langle A_{f,i}\rangle. \tag{77}$$

and for the Reynolds-stress equations

$$\alpha_f \rho_f \frac{d}{dt}\langle u_{f,i} u_{f,j}\rangle = -\frac{\partial}{\partial x_k}(\alpha_f\,\rho_f\langle u_{f,i} u_{f,j} u_{f,k}\rangle) - \alpha_f\rho_f\langle u_{f,i} u_{f,k}\rangle\frac{\partial\langle U_{f,j}\rangle}{\partial x_k}$$

$$-\alpha_f\rho_f\langle u_{f,j} u_{f,k}\rangle\frac{\partial\langle U_{f,i}\rangle}{\partial x_k} + \alpha_f\,\rho_f\langle A_{f,i}\,v_{f,j} + A_{f,j}\,v_{f,i}\rangle + \alpha_f\,\rho_f\langle(B_f B_f^T)_{ij}\rangle.$$

$$(78)$$

Field equations for the discrete phase

Here, we wish to derive the field equations for the discrete particles (the partial differential equations for the expected values of the variables attached to a discrete particle) for the following quantities: the mean discrete particle velocity $\langle U_{p,i}\rangle$, the second order velocity moment for discrete particles $\langle u_{p,i} u_{p,j}\rangle$, the mean of the fluid velocity seen $\langle U_{s,i}\rangle$, the second order velocity moment for the fluid seen $\langle u_{s,i} u_{s,j}\rangle$, the fluid seen-discrete particle velocity correlation tensor, $\langle u_{s,i} u_{p,j}\rangle$, the mean diameter $\langle d_p\rangle$, the diameter-fluid velocity seen correlation vector, $\langle d_p' u_{s,i}\rangle$, the diameter-particle velocity correlation vector, $\langle d_p' u_{p,i}\rangle$ and the diameter second order moment $\langle (d_p')^2\rangle$.

This procedure (Chapman and Cowling, 1970; Liboff, 1998) is general and allows to calculate any moment, but as the present task is to derive an Eulerian-like model, closure is performed at the second order level. It is straightforward to prove that $F_p^E(t, \mathbf{x}; \mathbf{V}_p, \mathbf{V}_s, \boldsymbol{\delta}_p)$ verifies the following partial differential equation,

$$\frac{\partial F_p^E}{\partial t} + V_{p,i}\frac{\partial F_p^E}{\partial x_i} = -\frac{\partial}{\partial V_{p,i}}(A_{p,i}\,F_p^E)$$

$$-\frac{\partial}{\partial V_{s,i}}(A_{s,i}\,F_p^E) + \frac{1}{2}\frac{\partial^2}{\partial V_{s,i}\partial V_{s,j}}\left((B_s B_s^T)_{ij}\,F_p^E\right).$$

$$(79)$$

Like for the fluid phase, using Eq. (66), one can write

$$\frac{\partial}{\partial t}(\alpha_p\rho_p\langle \mathcal{H}_p\rangle) + \frac{\partial}{\partial x_i}(\alpha_p\rho_p\langle V_{p,i}\,\mathcal{H}_p\rangle) = \alpha_p\rho_p\langle A_{p,i}\frac{\partial\mathcal{H}_p}{\partial V_{p,i}}\rangle +$$

$$\alpha_p\rho_p\langle A_{s,i}\frac{\partial\mathcal{H}_p}{\partial V_{s,i}}\rangle + \alpha_p\rho_p\langle(B_s B_s^T)_{ij}\frac{\partial^2\mathcal{H}_p}{\partial V_{s,i}\partial V_{s,j}}\rangle.$$

$$(80)$$

The partial differential equations for the specified discrete particle expectations can now be derived, simply by choosing the right function for $\mathcal{H}_p$. For $\mathcal{H}_p = 1$, the continuity equation is obtained,

$$\frac{\partial}{\partial t}(\alpha_p\rho_p) + \frac{\partial}{\partial x_i}(\alpha_p\,\rho_p\langle U_{p,i}\rangle) = 0.$$

$$(81)$$

With $\mathcal{H}_p = V_{p,i}$, the momentum equation for the discrete phase reads (where the Eulerian derivative along the path of a discrete particle is denoted d/dt with $d/dt = \partial/\partial t + \langle U_{p,m}\rangle \partial/\partial x_m$)

$$\alpha_p \, \rho_p \frac{d}{dt}\langle U_{p,i}\rangle = -\frac{\partial}{\partial x_j}(\alpha_p \, \rho_p \langle u_{p,i} u_{p,j}\rangle) + \alpha_p \, \rho_p \langle A_{p,i}\rangle. \tag{82}$$

The partial differential equation of the expected fluid velocity seen, $\langle U_{s,i}\rangle$, is derived with $\mathcal{H}_p = V_{s,i}$

$$\alpha_p \, \rho_p \frac{d}{dt}\langle U_{s,i}\rangle = -\frac{\partial}{\partial x_j}(\alpha_p \, \rho_p \langle u_{s,i} u_{p,j}\rangle) + \alpha_p \, \rho_p \langle A_{s,i}\rangle, \tag{83}$$

and with $\mathcal{H}_p = \delta_p$, the partial differential equation for the mean diameter reads

$$\alpha_p \, \rho_p \frac{d}{dt}\langle d_p\rangle = -\frac{\partial}{\partial x_i}(\alpha_p \, \rho_p \langle d_p^{'} u_{p,i}\rangle). \tag{84}$$

The partial differential equations verified by the second order moments ($n+m+q = 2$ in Eq. (68)) can not be obtained directly from the procedure presented above. A change of coordinates in sample space is introduced, $\mathbf{v}_p = \mathbf{V}_p - \langle \mathbf{U}_p\rangle(t,\mathbf{x})$, $\mathbf{v}_s = \mathbf{V}_s - \langle \mathbf{U}_s\rangle(t,\mathbf{x})$ and $\delta_p^{'} = \delta_p - \langle d_p\rangle(t,\mathbf{x})$. It is straightforward to prove that the partial differential equation verified by $F_p^E(t,\mathbf{x};\mathbf{v}_p,\mathbf{v}_s,\delta_p^{'})$ reads

$$\frac{dF_p^E}{dt} + v_{p,i}\frac{\partial F_p^E}{\partial x_i} = -\frac{\partial}{\partial v_{p,i}}(A_{p,i}\, F_p^E) - \frac{\partial}{\partial v_{s,i}}(A_{s,i}\, F_p^E)$$

$$+ \frac{1}{2}\frac{\partial^2}{\partial v_{s,i} v_{s,j}}\left((B_s B_s^T)_{ij}\, F_p^E\right)$$

$$+ \frac{d\langle U_{p,i}\rangle}{dt}\frac{\partial F_p^E}{\partial v_{p,i}} + \frac{d\langle U_{s,i}\rangle}{dt}\frac{\partial F_p^E}{\partial v_{s,i}} + \frac{d\langle d_p\rangle}{dt}\frac{\partial F_p^E}{\partial \delta_p^{'}}$$

$$+ v_{p,i}\frac{\partial\langle U_{p,j}\rangle}{\partial x_i}\frac{\partial F_p^E}{\partial v_{p,j}} + v_{p,i}\frac{\partial\langle U_{s,j}\rangle}{\partial x_i}\frac{\partial F_p^E}{\partial v_{s,j}} + v_{p,i}\frac{\partial\langle d_p\rangle}{\partial x_i}\frac{\partial F_p^E}{\partial \delta_p^{'}}. \tag{85}$$

With $F_p^E(t,\mathbf{x};\mathbf{v}_p,\mathbf{v}_s,\delta_p^{'})\,d\mathbf{v}_p\,d\mathbf{v}_s\,d\delta_p^{'} = F_p^E(t,\mathbf{x};\mathbf{V}_p,\boldsymbol{\psi}_p)\,d\mathbf{V}_p\,d\boldsymbol{\psi}_p$, because we have

$$\left|\frac{\partial(\mathbf{V}_f,\mathbf{V}_p,\mathbf{V}_s,\boldsymbol{\delta}_p)}{\partial(\mathbf{v}_f,\mathbf{v}_p,\mathbf{v}_s,\boldsymbol{\delta}_p^{'})}\right| = 1, \tag{86}$$

the moments of order $n+m+q$ can also be defined with $F_p^E(t,\mathbf{x};\mathbf{v}_p,\mathbf{v}_s,\delta_p)$. The partial differential equation for a function $\mathcal{H}_p(\mathbf{v}_p,\mathbf{v}_s,\delta_p)$ is derived in the same fashion as for Eq. (80). Once again, assuming that all generalized integrals converge and that uniform convergence is verified, after some derivations, the partial differential equation verified by $\mathcal{H}_p(\mathbf{v}_p, \mathbf{v}_s, \delta_p')$ becomes

$$\frac{d}{dt}(\alpha_p \rho_p \langle \mathcal{H}_p \rangle) + \frac{\partial}{\partial x_i}(\alpha_p \rho_p \langle v_{p,i} \mathcal{H}_p \rangle) =$$

$$\alpha_p \rho_p \langle A_{p,i} \frac{\partial \mathcal{H}_p}{\partial v_{p,i}} \rangle + \alpha_p \rho_p \langle A_{s,i} \frac{\partial \mathcal{H}_p}{\partial v_{s,i}} \rangle + \frac{1}{2} \alpha_p \rho_p \langle (B_s B_s^T)_{ij} \frac{\partial^2 \mathcal{H}_p}{\partial v_{s,i} \partial v_{s,j}} \rangle$$

$$- \alpha_p \rho_p \frac{d \langle U_{p,i} \rangle}{dt} \langle \frac{\partial \mathcal{H}_p}{\partial v_{p,i}} \rangle - \alpha_p \rho_p \frac{d \langle U_{s,i} \rangle}{dt} \langle \frac{\partial \mathcal{H}_p}{\partial v_{s,i}} \rangle - \alpha_p \rho_p \frac{d \langle d_p \rangle}{dt} \langle \frac{\partial \mathcal{H}_p}{\partial \delta_p'} \rangle$$

$$- \alpha_p \rho_p \frac{\partial \langle U_{p,j} \rangle}{\partial x_i} \langle \frac{\partial (v_{p,i} \mathcal{H}_p)}{\partial v_{p,j}} \rangle - \alpha_p \rho_p \frac{\partial \langle U_{s,j} \rangle}{\partial x_i} \langle \frac{\partial (v_{p,i} \mathcal{H}_p)}{\partial v_{s,j}} \rangle$$

$$- \alpha_p \rho_p \frac{\partial \langle d_p \rangle}{\partial x_i} \langle \frac{\partial (v_{p,i} \mathcal{H}_p)}{\partial \delta_p'} \rangle.$$

$$(87)$$

The partial differential equations for the velocity moments of order 2 can now be obtained. Inserting $\mathcal{H}_p = v_{p,i} v_{p,j}$, $\mathcal{H}_p = v_{s,i} v_{p,j}$ and $\mathcal{H}_p = v_{s,i} v_{s,j}$ in Eq. (87), the partial differential equations verified by the second order velocity moment for the discrete particles $\langle u_{p,i} u_{p,j} \rangle$, for the second order velocity moment of the fluid seen $\langle u_{s,i} u_{s,j} \rangle$ and for the fluid seen-discrete particle velocity correlation tensor $\langle u_{s,i} u_{p,j} \rangle$, can be derived. After some algebra, one finds for $\langle u_{p,i} u_{p,j} \rangle$

$$\alpha_p \rho_p \frac{d}{dt} \langle u_{p,i} u_{p,j} \rangle = -\frac{\partial}{\partial x_k}(\alpha_p \rho_p \langle u_{p,i} u_{p,j} u_{p,k} \rangle) - \alpha_p \rho_p \langle u_{p,i} u_{p,k} \rangle \frac{\partial \langle U_{p,j} \rangle}{\partial x_k}$$

$$- \alpha_p \rho_p \langle u_{p,j} u_{p,k} \rangle \frac{\partial \langle U_{p,i} \rangle}{\partial x_k} + \alpha_p \rho_p \langle A_{p,i} v_{p,j} + A_{p,j} v_{p,i} \rangle,$$

$$(88)$$

for $\langle u_{p,i} u_{s,j} \rangle_p$

$$\alpha_p \rho_p \frac{d}{dt} \langle u_{s,i} u_{p,j} \rangle = -\frac{\partial}{\partial x_k}(\alpha_p \rho_p \langle u_{s,i} u_{p,j} u_{p,k} \rangle) - \alpha_p \rho_p \langle u_{s,i} u_{p,k} \rangle \frac{\partial \langle U_{p,j} \rangle}{\partial x_k}$$

$$- \alpha_p \rho_p \langle u_{p,j} u_{p,k} \rangle \frac{\partial \langle U_{s,i} \rangle}{\partial x_k} + \alpha_p \rho_p \langle A_{s,i} v_{p,j} \rangle + \alpha_p \rho_p \langle A_{p,j} v_{s,i} \rangle,$$

$$(89)$$

and for $\langle u_{s,i} u_{s,j} \rangle_p$

$$\alpha_p \rho_p \frac{d}{dt} \langle u_{s,i} u_{s,j} \rangle = - \frac{\partial}{\partial x_k} (\alpha_p \, \rho_p \langle u_{s,i} u_{s,j} u_{s,k} \rangle) - \alpha_p \rho_p \langle u_{s,i} u_{s,k} \rangle \frac{\partial \langle U_{s,j} \rangle}{\partial x_k}$$

$$- \alpha_p \rho_p \langle u_{s,j} u_{s,k} \rangle \frac{\partial \langle U_{s,i} \rangle}{\partial x_k} + \alpha_p \, \rho_p \langle A_{s,j} \, v_{s,i} + A_{s,i} \, v_{s,j} \rangle + \alpha_p \, \rho_p \langle (B_s B_s^T)_{ij} \rangle.$$

$$(90)$$

By replacing $\mathcal{H}_p$ by $\mathcal{H}_p = \delta_p' v_{p,i}$, $\mathcal{H}_p = \delta_p' v_{s,i}$ and $\mathcal{H}_p = (\delta_p')^2$ in Eq. (87), the partial differential equations verified by the second order discrete particle velocity-diameter moment $\langle d_p' u_{p,i} \rangle$, by the second order fluid velocity seen-diameter moment $\langle d_p' u_{s,i} \rangle$ and by the second order diameter moment can be written. After some calculus, one finds for $\langle d_p' u_{p,i} \rangle$

$$\alpha_p \rho_p \frac{d}{dt} \langle d_p' u_{p,i} \rangle = - \frac{\partial}{\partial x_j} (\alpha_p \, \rho_p \langle d_p' u_{p,i} u_{p,j} \rangle) - \alpha_p \rho_p \langle d_p' u_{p,j} \rangle \frac{\partial \langle U_{p,i} \rangle}{\partial x_j}$$

$$- \alpha_p \rho_p \langle u_{p,i} u_{p,j} \rangle \frac{\partial \langle d_p \rangle}{\partial x_j} + \alpha_p \, \rho_p \langle A_{p,i} \, d_p' \rangle,$$

$$(91)$$

for $\langle d_p' u_{s,i} \rangle$

$$\alpha_p \rho_p \frac{d}{dt} \langle d_p' u_{s,i} \rangle = - \frac{\partial}{\partial x_j} (\alpha_p \, \rho_p \langle d_p' u_{s,i} u_{p,j} \rangle) - \alpha_p \rho_p \langle d_p' u_{p,j} \rangle \frac{\partial \langle U_{s,i} \rangle}{\partial x_j}$$

$$- \alpha_p \rho_p \langle u_{s,i} u_{p,j} \rangle \frac{\partial \langle d_p \rangle}{\partial x_j} + \alpha_p \, \rho_p \langle A_{s,i} \, d_p' \rangle,$$

$$(92)$$

and for $\langle (d_p')^2 \rangle$

$$\alpha_p \rho_p \frac{d}{dt} \langle (d_p')^2 \rangle = - \frac{\partial}{\partial x_i} (\alpha_p \, \rho_p \langle (d_p')^2 u_{p,i} \rangle) - 2 \alpha_p \rho_p \langle d_p' u_{p,i} \rangle \frac{\partial \langle d_p \rangle}{\partial x_i}. \qquad (93)$$

Closure of the two-field model

Mean field equations have now been written, up to the second order moments, in the case of a non-reactive dispersed two-phase flow where the fluid is incompressible and the discrete particles are hard spheres of constant density. The set of equations and the associated unclosed terms are given in Table 1. Table 1 should convince the reader of the necessity of a Lagrangian approach when one attempts to simulate turbulent reactive dispersed two-phase flows: the physics of the problem are quite simplified and the system of equation is already nearly intractable: 13 partial differential equations (the dimension of the system is 46) with 23 terms to be closed.

Table 1. Two-field model: list of the mean field equations and related unclosed terms

variable	unclosed terms	third order tensors
α_f		
α_p		
$\langle U_f \rangle$	$\langle A_{f,i} \rangle \ \ I^M_{pf,i}$	
$\langle U_p \rangle$	$\langle A_{p,i} \rangle$	
$\langle U_s \rangle$	$\langle A_{s,i} \rangle \ \ \langle A^D_{p,i} \rangle$	
$\langle d_p \rangle$	$\langle d'_p u_{p,i} \rangle$	
$\langle u_{f,i} u_{f,j} \rangle$	$\langle A_{f,i} u_{f,j} \rangle \ \ I^R_{pf,ij} \ \ \langle (B_f B_f^T)_{ij} \rangle$	$\langle u_{f,i} u_{f,j} u_{f,k} \rangle$
$\langle u_{p,i} u_{p,j} \rangle$	$\langle A_{p,i} u_{p,j} \rangle$	$\langle u_{p,i} u_{p,j} u_{p,k} \rangle$
$\langle u_{s,i} u_{s,j} \rangle$	$\langle A_{s,i} u_{s,j} \rangle \ \ \langle A^D_{p,j} u_{s,i} \rangle \ \ \langle (B_s B_s^T)_{ij} \rangle$	$\langle u_{s,i} u_{s,j} u_{s,k} \rangle$
$\langle u_{s,i} u_{p,j} \rangle$	$\langle A_{s,i} u_{p,j} \rangle \ \ \langle A_{p,j} u_{s,i} \rangle \ \ \langle A^D_{p,j} u_{p,i} \rangle$	$\langle u_{s,i} u_{p,j} u_{p,k} \rangle$
$\langle d'_p u_{p,i} \rangle$	$\langle A_{p,i} d'_p \rangle$	$\langle d'_p u_{p,i} u_{p,j} \rangle$
$\langle d'_p u_{s,i} \rangle$	$\langle A_{s,i} d'_p \rangle \ \ \langle A^D_{p,i} d'_p \rangle$	$\langle d'_p u_{s,i} u_{p,j} \rangle$
$\langle (d'_p)^2 \rangle$		$\langle (d'_p)^2 u_{p,i} \rangle$

However, in the case of industrial applications which fall into the category of the simplified case under consideration (for example in fluidised beds), the mean field field equations can be furthermore simplified, provided some additional restrictions and hypotheses. In this case, the model consists in partial differential equations for α_p, α_f, for the expected velocities $\langle U_f \rangle, \langle U_p \rangle, \langle U_s \rangle$ and the traces of the contracted tensors, that is the turbulent kinetic energies in both phases $\langle u_{f,i} u_{f,i} \rangle/2, \langle u_{p,i} u_{p,i} \rangle/2$ and the fluid-particle velocity covariance $\langle u_{s,i} u_{p,i} \rangle$. The treatment of the closures is out of the scope of the present paper and can be found in Peirano and Leckner (1998). An interesting treatment of the issue of moments is discussed in the chapter written by R Fox.

3.3   State-of-the-art Modelling

With the present choice of the state vector, the stochastic process used to describe the system has been chosen $\mathbf{Z} = (\mathbf{x}_p, \mathbf{U}_p, \mathbf{U}_s)$. Following the trajectory point of view, we must now propose a time-evolution equation for $\mathbf{U}_s$ that together with Eqs. (2) will give the complete SDE for the components of $\mathbf{Z}$. Compared to most Lagrangian models which are often built in a discrete setting, we are looking for a model written in continuous time to be consistent with our mathematical framework.

From the physical point of view, a time-evolution equation for $\mathbf{U}_s$ amounts to modelling turbulent dispersion, an issue which is more complicated than turbulent diffusion. Indeed, particle inertia (τ_p) and the effect of an external force field induce a separation of the fluid element and of the discrete particle initially located at the same point, as represented in Fig. 1. In the asymptotic limit of small particle inertia, $\tau_p \to 0$, and in absence of external forces this separation effect disappears and the problem of modelling diffusion is retrieved, for which the stochastic model discussed by Pope (1994) can be applied. For that reason, dispersion models (simulation of $\mathbf{U}_s$) are extensions of diffusion models (simulation of $\mathbf{U}_f$).

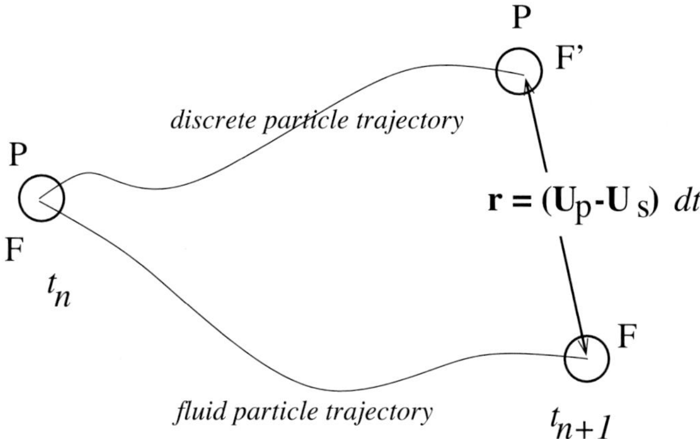

Figure 1. Fluid element and particle paths.

An extensive description of the physical aspects of turbulent dispersion can be found elsewhere (Pozorski and Minier, 1998; Minier and Peirano, 2001), so we recall just the key points that are used to device a stochastic model. It is proposed to consider separately the two physical effects of particle inertia and external forces. Two non-dimensional numbers have introduced for that purpose: particle inertia is measured by the Stoke number $St = \frac{\tau_p}{T_L}$, and external forces by $\xi = \frac{|V_r|}{u'}$, u' being a characteristic fluid turbulent velocity ($u' = \sqrt{2/3k}$). Then, we consider successively the influence of these two effects on the characteristics of $\mathbf{U}_s$:

(i) in the absence of external forces ($\xi = 0$), only particle inertia plays a role. We expect then that the characteristic, or integral, timescale of the velocity of the fluid seen, say $T_L^*(\xi = 0)$ varies between the fluid Lagrangian timescale T_L, in the limit of low St numbers, and the Eulerian timescale T_E in the limit of high St numbers.

(ii) Leaving out particle inertia, external forces creates mean drifts ($\xi \neq 0$)

and induce a decorrelation of the velocity of the fluid seen with respect to the velocity of fluid particles. This effect is called CTE (crossing trajectory effect) and is related to a mean relative velocity between particles and the fluid rather than an instantaneous one.

In the model presented here, we consider that T_E remains of the same order as T_L, which seems actually a reasonable choice since there is little information for complex flows. Detailed models have been proposed for particle inertia effect (Pozorski and Minier, 1998), and the present model can be easily extended accordingly. Here we consider that $T_L^*(\xi = 0) = T_L$. The representative picture is now sketched in Fig. 2 where only mean-drifts induce separation.

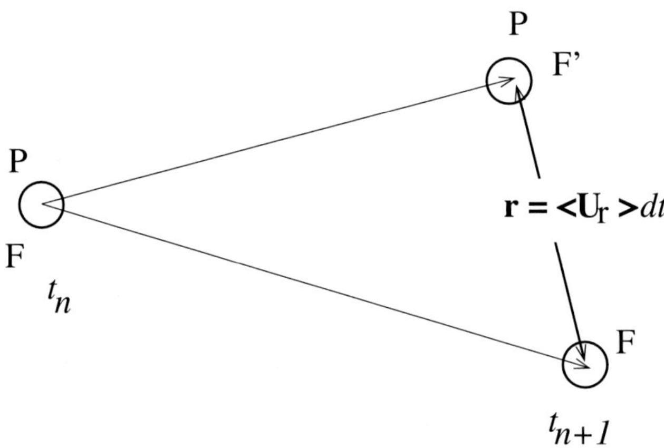

Figure 2. Mean fluid and particle paths.

Langevin equation model

Using the physical description of the CTE effect as due to a mean-drift (Fig. 2), we can now, as in single phase case, apply the Kolmogorov theory to suggest a dispersion model. Indeed, if we introduce $v(\tau, \mathbf{r}) = u_f(t_0 + \tau, \mathbf{x_0} + u(t_0, \mathbf{x_0})\tau + \mathbf{r}) - u_f(t_0, \mathbf{x_0})$, the fluid velocity field relative to the velocity of the fluid particle F at time t_n, that is with $u_f(t_0, \mathbf{x_0}) = u_s(t_0)$ Fig. 2, we can write that:

$$d\mathbf{U}_s = v(dt, \langle \mathbf{U}_r \rangle \, dt), \tag{94}$$

where $\langle \mathbf{U}_r \rangle = \langle \mathbf{U}_p \rangle - \langle \mathbf{U}_f \rangle$ is the mean relative velocity between the discrete particle and the surrounding fluid element. Then, the differential change, and so the Eulerian statistics, of the fluid velocity seen depends on $\langle \epsilon \rangle$ and ν, and on the mean drift due to the CTE effect, but not on the instantaneous particle or fluid velocities. Since it is the mean velocity $\mathbf{U}_r$ that appears in Eq. (94), the Kolmogorov theory can then be applied (Monin and Yaglom, 1975), to show that in high-Reynolds number flows and for a time increment dt that belongs to the inertial range, we have

$$\langle dU_{s,i} dU_{s,j} \rangle = D_{ij}(dt), \tag{95}$$

where the matrix D_{ij} is determined by the two scalars functions $D_{||}$ and $D_\perp$ through

$$D_{ij} = D_\perp \delta_{ij} + \left[D_{||} - D_\perp \right] r_i r_j. \tag{96}$$

The separation vector $\mathbf{r}$ being in the direction of the mean relative velocity $\mathbf{r} = \langle \mathbf{U}_r \rangle / |\langle \mathbf{U}_r \rangle|$. Being the functions $D_{||}$ and $D_\perp$ respectively the aligned and transverse velocity correlation. Dimensional analysis yields that in the inertial range, we have

$$D_{||}(dt) = \langle \epsilon \rangle \, dt \, \alpha_{||} \left(\frac{|\langle \mathbf{U}_r \rangle|^2}{\langle \epsilon \rangle dt} \right), \qquad D_\perp(dt) = \langle \epsilon \rangle \, dt \, \alpha_\perp \left(\frac{|\langle \mathbf{U}_r \rangle|^2}{\langle \epsilon \rangle dt} \right). \tag{97}$$

For the two function $\alpha_{||}$ and $\alpha_\perp$ there is no exact prediction, but in two limit cases they can be explicitely computed. When the mean relative velocity is small, $|\langle \mathbf{U}_r \rangle| \ll (\langle \epsilon \rangle dt)^{1/2}$ for a given time interval dt, the statistics of the velocity of the fluid seen are expected to be close to the fluid ones, and thus

$$\frac{|\langle \mathbf{U}_r \rangle|^2}{\langle \epsilon \rangle dt} \ll 1, \quad \Rightarrow \alpha_{||} \simeq \alpha_\perp \simeq C_0. \tag{98}$$

On the other hand, when the relative mean velocity is large ($|\langle \mathbf{U}_r \rangle| \gg (\langle \epsilon \rangle dt)^{1/2}$), we can resort to the frozen turbulence hypothesis. In that case, we obtain that

$$D_{||}(dt) \simeq C(\langle \epsilon \rangle \langle \mathbf{U}_r \rangle \, dt)^{2/3}, \qquad D_\perp(dt) \simeq \frac{4}{3} C(\langle \epsilon \rangle \langle \mathbf{U}_r \rangle \, dt)^{2/3}, \tag{99}$$

which shows that, in that limit, the two functions $\alpha_{||}(x)$ and $\alpha_\perp(x)$ vary as $x^{1/3}$. Some comments are in order. The same procedure applied to fluid phase leads to a time correlation which is linear in time, that which supports a diffusion model (Monin and Yaglom, 1975; Pope, 1994). In the particle case, this remains true for small relative velocity, whereas a different behaviour is predicted in the limit of frozen turbulence.

Nevertheless, a useful approximation can be proposed. Indeed, if we freeze the values of the functions $\alpha_{||}$ and $\alpha_\perp$ for a certain value of the time interval, say Δt_r and write

$$D_{||}(dt) \simeq \langle \epsilon \rangle \, dt \, \alpha_{||} \left(\frac{|\langle \mathbf{U}_r \rangle|^2}{\langle \epsilon \rangle \Delta t_r} \right), \qquad D_{\perp}(dt) \simeq \langle \epsilon \rangle \, dt \, \alpha_{\perp} \left(\frac{|\langle \mathbf{U}_r \rangle|^2}{\langle \epsilon \rangle \Delta t_r} \right),$$

$$(100)$$

we have now a linear variation of $D_{||}(dt)$ and $D_{\perp}(dt)$ with respect to the time interval dt. The reference time lag may be the Lagrangian timescale which is the timescale over which fluid velocities are correlated. And since $\langle \epsilon \rangle T_L \simeq k$, we have

$$D_{||}(dt) \simeq \langle \epsilon \rangle \, dt \, \alpha_{||} \left(\frac{|\langle \mathbf{U}_r \rangle|^2}{k} \right), \qquad D_{\perp}(dt) \simeq \langle \epsilon \rangle \, dt \, \alpha_{\perp} \left(\frac{|\langle \mathbf{U}_r \rangle|^2}{k} \right).$$

$$(101)$$

This result suggests now a Langevin equation model which consists in simulating $\mathbf{U}_s$ as a diffusion process. With respect to fluid case, a supplementary hypothesis has been made. As a consequence, the Langevin model does not yield the correct spectrum in the limit of large relative velocity or frozen turbulence. However, it is important to underline that this is not a much relevant feature for engineering purposes, where the macroscopic behaviour is the real subject of interest, the important properties are the integral time scales rather than the precise form of the spectrum. Thus, present Langevin models are not affected in practice by this limit. That has been largely shown by numerical simulations, as it will be seen soon. It is also clear that much work remains to be done to improve present stochastic models.

The general stochastic differential equations for the velocity seen process have the form

$$dU_{s,i} = A_{s,i}(t, \mathbf{Z}) \, dt + B_{s,ij}(t, \mathbf{Z}) \, dW_j \qquad (102)$$

where the drift vector $\mathbf{A}_s$ and the diffusion matrix $\mathbf{B}_s$ have to be modelled.

Now, referring for all details to reach this form to Minier and Peirano (2001), we can propose the complete form of the stochastic model:

$$\begin{aligned} dU_{s,i} \;=\; & -\frac{1}{\rho_f} \frac{\partial \langle P \rangle}{\partial x_i} \, dt + \left(\langle U_{p,j} \rangle - \langle U_{f,j} \rangle \right) \frac{\partial \langle U_{f,i} \rangle}{\partial x_j} \, dt \\ & -\frac{1}{T_{L,i}^*} \left(U_{s,i} - \langle U_{f,i} \rangle \right) dt \\ & + \sqrt{ \langle \epsilon \rangle \left(C_0 b_i \tilde{k}/k + \frac{2}{3}(b_i \tilde{k}/k - 1) \right) } \, dW_i . \end{aligned}$$

$$(103)$$

The CTE has been modelled with the changing of timescales in drift and diffusion terms according to Csanady's analysis. Assuming for the sake of simplicity that the mean drift is aligned with the first coordinate axis, the modelled expressions for the timescales are, in the longitudinal direction:

$$T_{L,1}^* = \frac{T_L}{\sqrt{1 + \beta \dfrac{|\langle \mathbf{U}_r \rangle|^2}{2k/3}}} \tag{104}$$

and in the transversal directions (axis labeled 2 and 3)

$$T_{L,2}^* = T_{L,3}^* = \frac{T_L}{\sqrt{1 + 4\beta \dfrac{|\langle \mathbf{U}_r \rangle|^2}{2k/3}}} \tag{105}$$

where β is the ratio of the Lagrangian and the Eulerian timescales of the fluid $\beta = T_L/T_E$. In the diffusion matrix we have introduced a new kinetic energy:

$$b_i \frac{T_L}{T_L^*} \;\; ; \;\; \tilde{k} = \frac{3}{2} \frac{\sum_{i=1}^3 b_i \langle u_{f,i}^2 \rangle}{\sum_{i=1}^3 b_i}. \tag{106}$$

The case of general axis direction is discussed in Minier and Peirano (2001).

As a final remark, in the absence of mean drifts, the stochastic model for $\mathbf{U}_s$ reverts to the Langevin equation model used in single phase PDF modelling (Pope, 1994) and is thus free of any spurious drift by construction.

Equivalence with PDF approach

According to the arguments developed in the first chapter, the complete Langevin equation model (for the state vector $\mathbf{Z} = (\mathbf{x}_p, \mathbf{U}_f, \mathbf{U}_s)$) can be written

$$dx_{p,i} = U_{p,i}\, dt \tag{107a}$$

$$dU_{p,i} = \frac{1}{\tau_p}\left(U_{s,i} - U_{p,i}\right) dt + g_i\, dt \tag{107b}$$

$$dU_{s,i} = [A_{s,i}(t, \mathbf{Z}) + A_{p \to s,i}(t, \mathbf{Z}, \langle \mathbf{Z} \rangle)]\, dt + B_{s,ij}(t, \mathbf{Z})\, dW_j. \tag{107c}$$

This formulation is equivalent to a Fokker-Planck equation given in closed form for the corresponding pdf $p(t; \mathbf{y}_p, \mathbf{V}_p, \mathbf{V}_s)$ which is, in sample space

$$\frac{\partial p_p^L}{\partial t} + V_{p,i} \frac{\partial p_p^L}{\partial y_{p,i}} = -\frac{\partial}{\partial V_{p,i}}(A_{p,i}\, p_p^L)$$

$$- \frac{\partial}{\partial V_{s,i}}([A_{s,i} + \langle A_{p \to s,i} | \mathbf{y}_p, \mathbf{V}_p, \mathbf{V}_s \rangle]\, p_p^L) + \frac{1}{2} \frac{\partial^2}{\partial V_{p,i} \partial V_{p,j}}([B_p B_p^T]_{ij}\, p_p^L). \tag{108}$$

3.4   Numerical example: the Hercule experiment

The numerical simulation of this interesting test-case has been presented in detail elsewhere (Minier et al., 2004), and we only report here the main results. The theoretical model developed previously represents a PDF model for the particle phase only. It does not contain any description of the continuous or fluid phase. As we have explained in previous sections, a full Lagrangian formalism and model is available and, thus, it is in principle possible to extend the PDF description to both fluid and particle phases. However, at the moment, this complete PDF approach is limited for practical calculations, and, in the present work, a classical second-moment approach is followed for the continuous phase. The complete numerical model is therefore a hybrid method and corresponds to a classical approach referred to as Eulerian/Lagrangian in literature. The terminology is not actually adequate to describe the complete model, and it would be better to talk of a Moment/pdf hybrid model, but corresponds to the numerical approach. Indeed, from the numerical point of view, the fluid phase is modelled by mean fluid, obtained by solving partial differential equations on a grid with an Eulerian approach, while the particle phase is modelled by a large number of Lagrangian particles distributed in the domain and whose properties are obtained by solving stochastic differential equations. It is worth emphasizing that these particles are now stochastic particles, or more precisely samples of the underlying pdf, rather than precise models of the actual particles. The overall numerical method is therefore an example of Monte Carlo particle/mesh technique.

In particular, the stochastic modelling together with the hybrid nature of the global approach raise specific issues. Notably, it is very important not to overlook consistency issues (Chibbaro and Minier, 2011a). Although a detailed analysis of the numerical approach is out of the present scope, it is important to retain that the numerical approach involves many issues (Hockney and Eastwood, 1988; Peirano et al., 2006), which have not been always investigated or may have been overlooked, such as consistent discrete averages and mass-continity constraint.

General Algorithm

The flow-chart of the code is shown in fig. 3. At each time step, the fluid mean fields are first computed by solving the corresponding partial differential equations (RSM model) with a classical finite volume aproach. The Eulerian solver then provides the Lagrangian solver with the mean fields that are necessary to advance particles properties. Within the Lagrangian solver, the dispersed phase is represented by a large number of particles

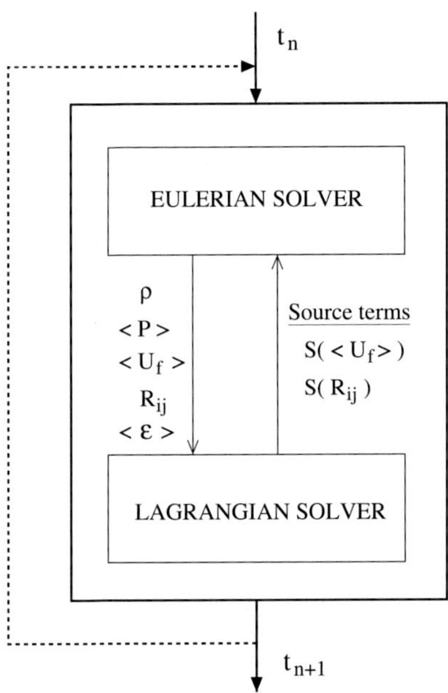

Figure 3. Sketch of the algorithm for one time step.

and, as proposed by the model, the state vector attached to each particle is $\mathbf{Z} = (\mathbf{x}_p, \mathbf{U}_f, \mathbf{U}_s)$. Once particle properties have been updated, particles are located in the grid cells used also for the fluid computation, and mean statistical properties, such as the mean particle velocity field $\langle \mathbf{U}_p \rangle$, are extracted by calculating ensemble averages from the set of particles present in each cell. In the case of two-way coupling, where particles modify the fluid flow, source terms accounting for momentum and energy exchange between the two phases are also calculated and are fed back into the Eulerian solver for the next time step computation. It is then seen that the two solvers are only loosely-coupled. This may lead to numerical difficulties when the particle loading is increased, consequently the source terms become important and the equations stiff. However, our present aim is to model moderate particle loading phenomena, and particle-particle collisions have been neglected. In this range, particles can still modify the fluid flow in a noticeable way but source terms remain low enough so that the loosely-coupled algorithm works well.

For stationary flows, such as the one considered later on, ensemble averages computed in every cell are then averaged in time, once the stationary regime has been reached. This time-averaging procedure is very helpful to reduce statistical noise to a negligible level (Muradoglu et al., 1999; Jenny et al., 2001; Muradoglu et al., 2001).

Within the particle solver, the particle properties are modelled by a vectorial SDE that we can write as

$$dZ_i = A_i(t, \mathbf{Z}), \langle f(\mathbf{Z}) \rangle, \langle \mathbf{X} \rangle)dt + B_{ij}(t, \mathbf{Z}, \langle f(\mathbf{Z}) \rangle, \langle \mathbf{X} \rangle)dW_j \qquad (109)$$

where f is a general function depending on the model and $\mathbf{X}$ stands for fluid fields. It is worth emphasizing that the drift and diffusion depend on statistics derived from the pdf that is implicitly calculated. Therefore, these SDE are different from standard ones (Arnold, 1974; Talay, 1995). Updating particles properties implies three steps:

1. projection of $\langle f(\mathbf{Z}) \rangle$ and $\langle \mathbf{X} \rangle$ at particle positions,
2. time integration of eq. (109),
3. averaging to compute the new values of $\langle f(\mathbf{Z}) \rangle$.

These steps involve many issues, whose discussion is out of the present scope (Kloeden and Platen, 1992; Talay, 1995; Minier et al., 2003; Peirano et al., 2006; Hockney and Eastwood, 1988)

Experimental setup

The experimental setup is typical for pulverized coal combustion where primary air and coal are injected in the center and secondary air is introduced on the periphery, Figure 4.

The experimental setup is typical for pulverised coal combustion where primary air and coal are injected in the centre and secondary air is introduced on the periphery, Fig. 4a. This is a typical bluff-body flow where the gas (air at ambient temperature, $T = 293\,K$) is injected in the inner region and also in the outer region where the inlet velocity is high enough to create a recirculation zone downstream of the injection. Solid particles (glass particles of density $\rho_p = 2450\,kg/m^3$) are then injected from the inner cylinder with a given mass flow rate and from there interact with the gas turbulence. This is a coupled turbulent two-phase flow since the particle mass loading at the inlet is high enough (22%) for the particles to modify the fluid mean velocities and kinetic energy. This is also a polydispersed flow where particle diameters vary according to a known distribution at the inlet, typically between $d_p = 20\mu m$ and $d_p = 110\mu m$ around an average of $d_p \sim 60\,\mu m$. Further details on the experimental setup and the measurement techniques can be found in (Borée et al., 2001). This test-case has been also discussed elsewhere (Minier et al., 2004). In figure 5a, we show the effect of the choice of the turbulence model for fluid velocity. For fluid phase, as previsible, the

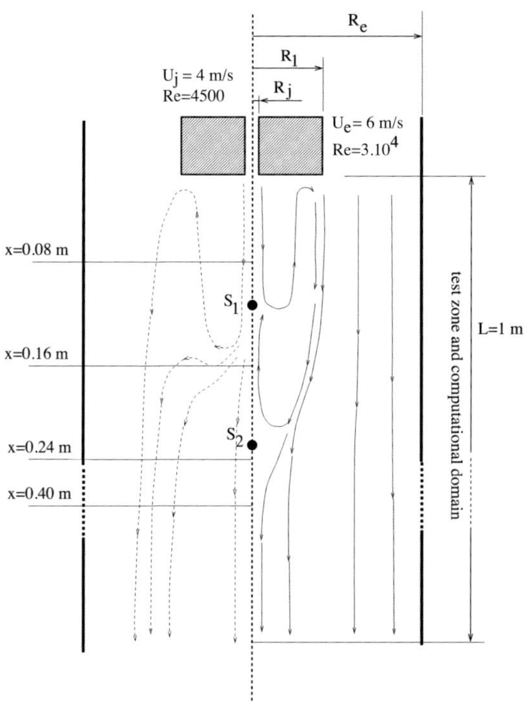

Figure 4. The 'Hercule' experimental setup: two stagnation points Experimental data are available for radial profiles of different statistical quantities at five axial distances downstream of the injection ($x = 0.08, 0.16, 0.24, 0.32, 0.40m$)

recirculation point is much better captured with a Reynolds stress model. For this reason, all the following calculations are performed with this fluid turbulence model. It is worth saying that there are possible consistency issues in hybrid Eulerian/Lagrangian (Chibbaro and Minier, 2011a). The Reynolds stress model used is consistent with the Lagrangian model and thus assures the global consistency of the method. For the particle phase, we show in figure 5b the mean axial velocity for two Lagrangian models. It is possible to appreciate that the CTE included in the mean gradient term in eq. (103) plays a role. The agreement with experimental results is fair but not perfect. In figure 6, the fluctuating particle velocity is shown along the radial direction at different positions on the axis. Typically this quantity is

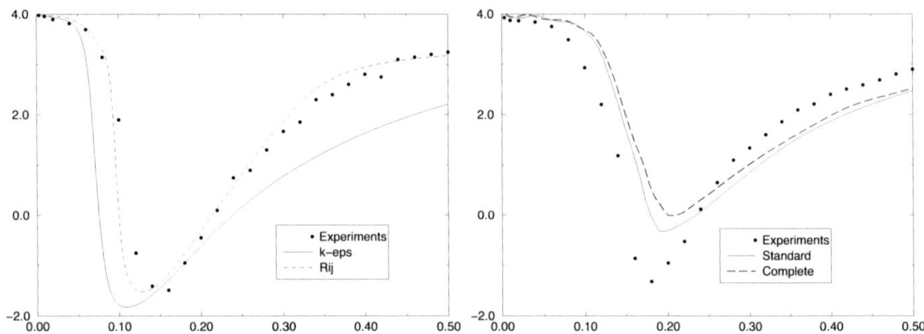

Figure 5. (a)Mean axial fluid velocity computed through $k - \epsilon$ and R_{ij} model. (b) Mean axial particle velocity computed coupling with a R_{ij} turbulence fluid model. Standard model neglects the gradient correction in the model, which is instead retained in the complete one.

not accessible through standard two-fluid simulations and is of much interest. Considering that this test-case is particularly challenging and that the radial velocity appears as the most difficult to predict (even the experiments show very noisy data), the results obtained are quite good. The qualitative behaviour is always captured and, at least at some positions, profiles are recovered quantitatively. Moreover, we can use the present numerical simulation to outline the kind of information that is available. As already indicated, in a number of engineering applications and notably in sediment flows, one is usually interested in having far more information than simply one or two moments. One of the interests of present stochastic models is to provide such information. For example, one would like to know how much time particles found in a certain volume have actually spent in that volume, or even how many of the particles present have previously entered another specific zone. This is illustrated in Fig. 7 which presents a plot of the instantaneous locations of the particles that are simulated at that time step. In this plot, particles are coloured by their residence time which reveals the recirculation. The denser plot indicates the accumulation of particles in the recirculation zone. In the same figure, two distributions of particle residence time are extracted and shown at two locations. In a cell near the inlet, the distribution is highly peaked: most of the particles present in that cell have just been injected and their residence time is small contributing to the near delta-value close to the origin of time. A smaller number of particles are found with larger residence times: these are particles which

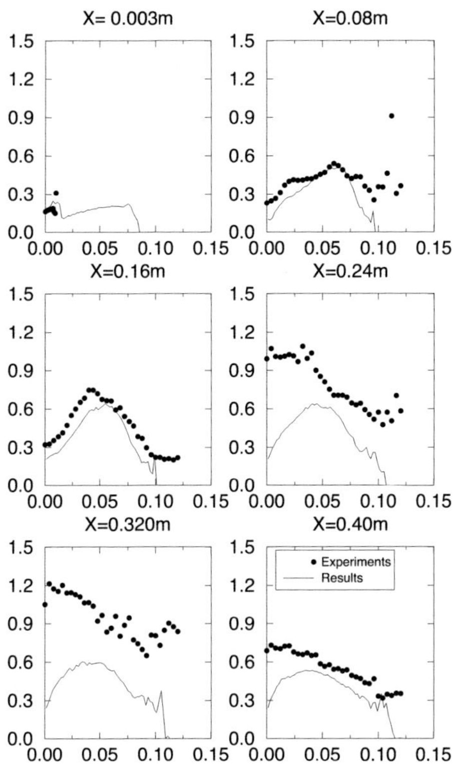

Figure 6. Fluctuating particle radial velocity

have recirculated and have gone back to the selected cell following different trajectories and thus having different residence times. A second distribution is also shown in Fig. 7, for a location near the outlet of the domain. In that case, we do not find different subclasses but rather a continuous spread of the particle residence time distribution. Particles have been well mixed since the injection and the distribution are smooth.

3.5   Conclusions and perspectives

In this chapter, we have applied a certain number of classical techniques of the stochastic approach to come up with a statistical description of turbulent polydispersed two-phase flows. We have named this approach mesoscopic, to distinguish it from the coarser RANS approach, and the much

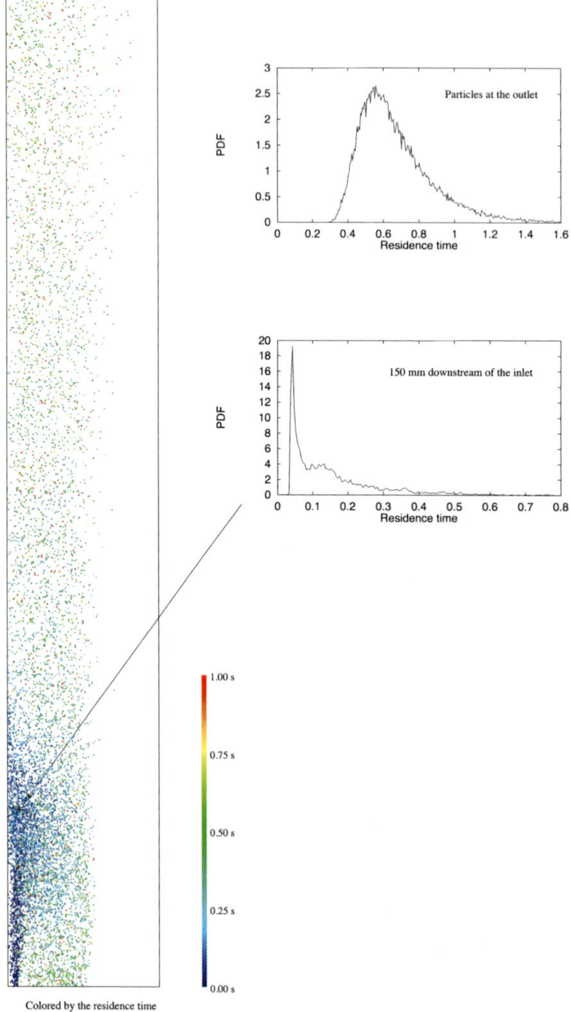

Figure 7. A snapshot of particle locations at one time step of the calculation where particles are coloured by their residence time within the domain. Two distributions of particle residence time are plotted at two different positions, near the injection and near the outlet.

finer DNS one. The stochastic model has been presented only after the general formalism has been put forward. This rigorous approach helps to develop in a safe manner the modelling step. Multiscale arguments have been used to make the choice of the state vector.

Some relevant specific points and perspectives have not been discussed or only marginally, for the sake of clarity. We want to point out some of them:

- The choice of the state vector for particle phase has been discussed at length. The choice made here is to keep in the state-vector the following variables: x_p, U_p, U_s. Historically, the first attempts to develop a PDF approach to polydispersed flows have been made gathering only x_p, U_p. This approach is usually called kinetic, in analogy with Boltzmann approach to gas dynamics. The kinetic approach has the merit to have advocated a PDF approach to polydispersed flows but is affected by many drawbacks which make it useless for simulations and not adequate for general non-homogeneous flows. In fact, it is not an actual PDF approach, since only moments equations derived from the kinetic equations can be used in practice. For these reasons, we consider the present choice x_p, U_p, U_s as the strict minimum for a consistent stochastic approach. On the contrary more general state vector could be considered to improve present Langevin models. Some elements of these problems can be found in literature (Minier and Pozorski, 1999; Minier and Peirano, 2001). A more detailed critical revision will be published soon.

- The stochastic model presented here, eq. (103) has not an isotropic diffusion coefficient. This is a very important point very often overlooked (Chibbaro and Minier, 2011b). Indeed, the isotropic form of the diffusion coefficient taken from single-phase models does not guarantee consistency with the simple decaying isotropic turbulence.

- Considering the perspectives, and with the nowadays computer performances, the present stochastic approach can be considered as the "standard" approach to collisionless particle-laden flows, with regard to two-fluid models. Instead, for more fundamental-based studies, its extension to a hybrid LES/PDF approach is certainly interesting and subject of present research (Pozorski and Apte, 2009; Chibbaro and Minier, 2011a; Bianco et al., 2012). Two-point fluid models can also be considered. Finally, a very promising route is the development of stochastic models within smooth-particle-hydrodynamics (SPH) framework. Given that this approach is full Lagrangian, it seems to offer the best field on which build up a consistent and efficient Lagrangian stochastic approach (Szewc et al., 2012).

Acknowledgements

We thank Angelo Vulpiani (Università di Roma "La Sapienza") in a special way for his fruitfull contribution to the redaction of this section, notably for the sections concerning multiscale approaches.

Bibliography

L. Arnold. *Stochastic Differential Equations: Theory and Applications*. Wiley, New-York, 1974.

R. Balescu. *Statistical dynamics: matter out of equilibrium*. Imperial College Press, London, 1997.

F. Bianco, S. Chibbaro, C. Marchioli, M. V. Salvetti, and A. Soldati. Intrinsic filtering errors of lagrangian particle tracking in les flow fields. *Phys Fluids*, 24(4):045103, 2012.

G. Boffetta and A. Vulpiani. *Probabilità in Fisica: Un'introduzione*. Springer, 2012.

M. Boivin, O. Simonin, and K. D. Squires. Direct numerical simulation of turbulence modulation by particles in isotropic turbulence. *J. Fluid Mech.*, 375:235–263, 1998.

J. Borée, T. Ishima, and I. Flour. The effect of mass loading and inter-particle collisions on the development of the polydispersed two-phase flow downstream of a confined bluff body. *Journal of Fluid Mechanics*, Jan 2001.

R. Brown. A brief account of microscopical observations made in the months of june, july, and august, 1827 on the particles contained in the pollen of plants; and on the general existence of active molecules in organic and inorganic bodies. *Philos. Mag.*, 4:161, 1828.

P. Castiglione, M. Falcioni, A. Lesne, and A. Vulpiani. *Chaos and coarse graining in statistical mechanics*. Cambridge University Press, 2008.

S. Chapman and T. G. Cowling. *The mathematical theory of non-uniform gases*. Cambridge Mathematical Library, Cambridge, 1970.

S. Chibbaro and J.-P. Minier. A note on the consistency of hybrid eulerian/lagrangian approach to multiphase flows. *International Journal of Multiphase Flow*, 37(3):293–297, 2011a.

S. Chibbaro and J.-P. Minier. The fdf or les/pdf method for turbulent two-phase flows. In *Journal of Physics: Conference Series*, volume 318, page 042049. IOP Publishing, 2011b.

R. Clift, J. R. Grace, and M. E. Weber. *Bubbles, Drops and Particles*. Academic Press. New York, 1978.

W. E and B. Engquist. Multiscale modeling and computation. *Notices of the AMS*, 50(9):1063–1070, 2003.

A. Einstein. *Investigations on the Theory of the Brownian Movement.* Dover Publications, 1956a.

A. Einstein. *Investigations on the Theory of the Brownian Movement.* Dover Publications, 1956b.

P. Espanol. Statistical mechanics of coarse-graining. *Novel Methods in Soft Matter Simulations,* pages 2256–2256, 2004.

C. W. Gardiner. *Handbook of Stochastic Methods for Physics, Chemistry and the Natural Sciences.* Springer-Verlag, Berlin, 2nd edition, 1990.

R. Gatignol. The Faxén formulae for a rigid particle in an unsteady non-uniform Stokes flow. *Journal de Mécanique Théorique et Appliquée,* 1 (2):143–160, 1983.

R. W. Hockney and J. W. Eastwood. *Computer simulations using particles.* Institute of Physics Publishing, Bristol and Philadelphia, 1988.

P. Jenny, S. B. Pope, M. Muradoglu, and D. A. Caughey. A hybrid algorithm for the joint pdf equation of turbulent reactive flows. *Journal of Computational Physics,* 166(2):218–252, 2001.

P. E. Kloeden and E. Platen. *Numerical solution of stochastic differential equations.* Springer-Verlag, Berlin, 1992.

P. Langevin. Sur la théorie du mouvement brownien. *CR Acad. Sci. Paris,* 146(530-533), 1908.

R. L. Liboff. *Kinetic theory: classical, quantum, and relativistic descriptions.* Prentice Hall Advanced Reference Series, London, 2nd edition, 1998.

T. S. Lundgren. Distribution functions in the statistical theory of turbulence. *Physics of Fluids,* 10(5):969–975, 1967.

C. Marchioli and A. Soldati. Mechanisms for particle transfer and segregation in a turbulent boundary layer. *Journal of Fluid Mechanics,* 468(-1): 283–315, 2002.

M.R. Maxey and J.J. Riley. Equation of motion for a small rigid sphere in a nonuniform flow. *Phys. Fluids,* 26(4):883–889, 1983.

J-P. Minier and E. Peirano. The pdf approach to polydispersed turbulent two-phase flows. *Physics Reports,* 352(1–3):1–214, 2001.

J-P. Minier and J. Pozorski. Wall boundary conditions in the pdf method and application to a turbulent channel flow. *Phys. Fluids,* 11:2632–2644, 1999.

J-P. Minier, E. Peirano, and S. Chibbaro. Weak first- and second-order numerical schemes for stochastic stochastic differential equations appearing in lagrangian two-phase flow modeling. *Monte Carlo Methods and Appl.,* 9(2):93–133, 2003.

J-P. Minier, E. Peirano, and S. Chibbaro. Pdf model based on langevin equation for polydispersed two-phase flows applied to a bluff body-body gas-solid flow. *Phys. Fluids,* 16(7):2419, 2004.

A. S. Monin and A. M. Yaglom. *Statistical Fluid Mechanics.* MIT Press, Cambridge, Mass, 1975.

M. Muradoglu, P. Jenny, S. B. Pope, and D. A. Caughey. A consistent hybrid finite volume/particle method for the pdf equations of turbulent reactive flows. *J. Comput. Phys.*, 154:342–371, 1999.

M. Muradoglu, S. B. Pope, and D. A. Caughey. The hybrid method for the pdf equations of turbulent reactive flows: consistency conditions and correction algorithms. *Journal of Computational Physics*, 172(2):841–878, 2001.

G. Pedrizzetti and E. A. Novikov. On markov modelling of turbulence. *Journal of Fluid Mechanics*, 280(1):69–93, 1994.

E. Peirano and B. Leckner. Fundamentals of turbulent gas-solid flows applied to circulating fluidized bed combustion. *Prog. Energy Combust. Sci.*, 24:259–296, 1998.

E. Peirano and J.-P. Minier. Probabilistic formalism and hierarchy of models for polydispersed turbulent two-phase flows. *Phys. Rev. E*, 65:046301, 2002.

E. Peirano, S. Chibbaro, J. Pozorski, and J.-P. Minier. Mean-field/pdf numerical approach for polydispersed turbulent two-phase flows. *Prog. En. Comb. Sci.*, 32(3):315, 2006.

S. B. Pope. Pdf methods for turbulent reactive flows. *Prog. Energy Combust. Sci.*, 11:119–192, 1985.

S. B. Pope. Lagrangian pdf methods for turbulent reactive flows. *Annu. Rev. Fluid Mech.*, 26:23–63, 1994.

J. Pozorski and S.V. Apte. Filtered particle tracking in isotropic turbulence and stochastic modeling of subgrid-scale dispersion. *International Journal of Multiphase Flow*, 35(2):118–128, 2009.

J. Pozorski and J-P. Minier. On the lagrangian turbulent dispersion models based o the langevin equation. *Int. J. Multiphase Flow*, 24:913–945, 1998.

K. Szewc, J. Pozorski, and J.-P. Minier. Simulations of single bubbles rising through viscous liquids using smoothed particle hydrodynamics. *International Journal of Multiphase Flow*, 2012.

D. Talay. *Simulation of Stochastic Differential Equation, in* Probabilistic Methods in Applied Physics. Springer-Verlag, Berlin, 1995.

F. Toschi and E. Bodenschatz. Lagrangian properties of particles in turbulence. *Annual Review of Fluid Mechanics*, 41:375–404, 2009.

N. G. Van Kampen. *Stochastic processes in physics and chemistry*, volume 1. North holland, 1992.

Quadrature-Based Moment Methods for Polydisperse Multiphase Flows

Rodney O. Fox [*†]

[*] Department of Chemical and Biological Engineering, Iowa State University, Ames, Iowa, USA

[†] Laboratoire EM2C-UPR CNRS 288, Ecole Centrale Paris, Châtenay-Malabry, France

Abstract We provide a brief introduction to quadrature-based moment methods that can be used to model polydisperse multiphase flows. A more detailed description can be found in Marchisio and Fox (2013). Our focus here is to introduce the reader to the principal topics and to provide insight into the numerial algorithms. An example application of gas-particle flow is used to illustrate the methods.

1 Introduction

Quadrature-based moment methods (QBMM) can be used to model polydisperse multiphase flows. The list of references provided at the end of the chapter contains a relatively complete list of the author's publications on this topic, including many applications to chemical engineering systems.

1.1 Polydisperse multiphase flows

By polydisperse multiphase flow, we mean that the flow has a continuous phase and at least one disperse phase consisting of entities of varying mass, volume, composition, etc. In general, we refer to the disperse-phase entities as *particles* and the collective behavior of all particles is referred to as the disperse phase. Such flows arise in many practical situations. Examples of industrial multiphase flows include fluid catalytic cracking, catalytic combustion, coal gasification, Fischer-Tropsch synthesis, chemical-looping processes, dust incineration, pneumatic transport and spray drying. Important aerospace applications include helicopter brown-out wherein sand and dust are lifted by the helicopter rotor causing reduced visibility while landing and extremely high risk of accidents, and propulsion systems such as spray combustion and rocket boosters. Examples of environmental polydisperse

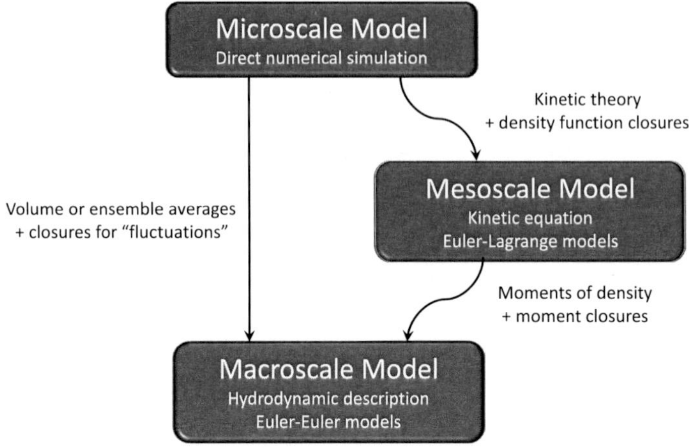

Figure 1. Modeling approach based on a kinetic equation at the mesoscale.

multiphase flows include gravity currents, volcanic eruptions, pollutant dispersion in the atmosphere (e.g., aerosols), and sand and dust storms.

The modeling challenges associated with polydisperse multiphase flows are numerous. For example, the volume occupied by the disperse phase can vary widely from point to point in the the same flow. In many cases, the particles will have an inertia very different from the continuous phase as measured by the wide range of Stokes numbers of the particles. Depending on the nature of the flow, the behavior of the particles can be dominated by collisions (or particle-particle interactions) or, in the opposite extreme, it can be collision-less with the two regimes coexisting in same flow. Also, depending on the properties of the particles, their instantaneous velocity fluctuations (or granular temperature) can be very small and very large in same flow. Finally, almost all disperse multiphase flows of practical interest are polydisperse (size, density, shape, etc.), which can have a strong effect on the fluid mechanical properties of the disperse phase. Thus, we will need a modeling framework that can handle all of the challenges that can occur simultaneously in the applications mentioned above.

1.2 Kinetic theory models

The modeling approach described in this chapter is based on using kinetic theory models for the disperse phase. As shown in Figure 1, a kinetic theory model is a mesoscale model that uses a kinetic equation to describe the properties of the particles. In contrast to a microscale model that contains a complete description of each individual particle and the flow of the continuous phase around the particles, the mesoscale model describes the evolution of a single particle coupled to the continuous phase through correlations such as for fluid drag on the particle or heat and mass transfer between phases, and interactions with other particles through a collision model (e.g., the Boltzmann hard-sphere collision integral). As shown in Figure 1 the mesoscale model is related to the macroscale model through the moments of the kinetic equation. For example, the zero-order moment leads to the diperse-phase continuity equation, while the first-order moments with respect to velocity lead to the disperse-phase momentum equation. It is important to recognize that the formal process of taking the moments of the kinetic equation results in a loss of information concerning the mesoscale properties of the flow (i.e., the macroscale model describes only a limited set of disperse-phase variables needed to represent "average" properties of the multiphase flow). Likewise, the mesoscale model contains much less information than the microscale model since, for example, two-particle correlations are not represented at the level of the one-particle kinetic equation.

The procedure used to develop a mesoscale model starting from the complete microscale description is shown schematically in Figure 2. By definition, the microscale model uses the complete set of equations needed to predict all relevant aspects of the multiphase flow (e.g., Navier-Stokes equation for the continuous phase along with boundary conditions at surface of each particle, Newton's second law for the momentum of each particle along with an exact description of particle collisions, etc.), but cannot be used to model a realistic application due, for example, to the large number of particles present. In order to find a tractable model, the set of independent variables must be drastically reduced to include only the mesoscale variables that describe a single particle. However, reducing the number of variables implies a loss of information that must be modeled in terms of the mesoscale variables. For example, the fluid forces exerted on the surface of each particle are exactly captured in the microscale description, but are modeled as an effective drag term that depends on the particle velocity and size in the mesoscale model. Likewise, for problems with heat and mass transfer, the microscale description must resolve the boundary layers around each particle and can predict the exact rates of exchange of heat and mass to/from each particle. In the mesoscale model, heat and mass transfer are modeled

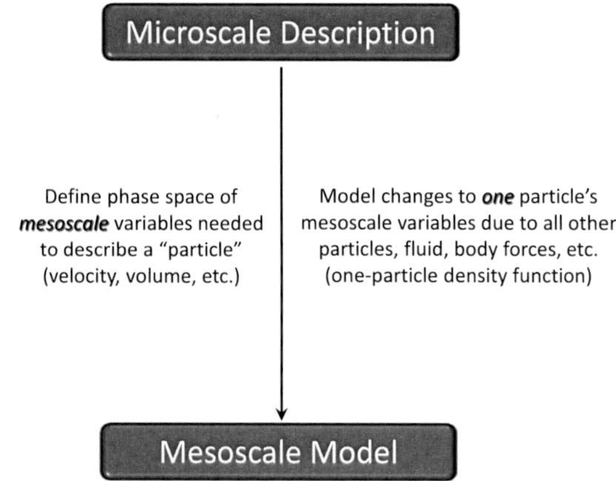

Figure 2. Passage from microscale description to the mesoscale model.

using a correlation expressing the rates as functions of the particle's temperature, concentration and velocity relative to the *average* local values in the fluid phase.

In kinetic theory, the space containing all possible values of the mesoscale variables is referred to as phase space. Depending on the number of mesoscale variables needed to describe a polydisperse multiphase flow, the dimension of phase space can be large. (We give some examples below.) In contrast, since the macroscale model is found by integrating over all of phase space, leaving only real space and time as the independent variables, the macroscale model uses a smaller set of independent variables. In summary, going from the microscale to the macroscale requires two levels of simplification. First, the complete microscale description is replaced with mesoscale variables that describe how each particle evolves through phase space, real space and time. In this step, the physics of the problem are simplified to retain only the quantities needed to represent the important properties of the flow. Second, information about individual particles is lost by integrating over the phase space containing all of the mesoscale variables, leaving only the *average* or *collective* properties of the disperse phase. This step results in unclosed terms in the transport equations for the moments that must be closed using only the knowledge available from the finite set of transported (or solved) moments. By definition, the solution to the kinetic equation contains a complete description of all moments. Thus, the moment-closure problem is mathematical (as opposed to physical as in the first step) because we need to represent the solution to the kinetic equation using a reduced set of information (i.e., weighted integrals over phase space). In comparison to classical methods used to derive the macroscale model directly from the microscale description wherein the physical and mathematical closures are mixed together in one step, using a mesoscale model allows for a clear separation between the two types of closure (Marchisio and Fox, 2013).

1.3   Example mesoscale models

There are many types of mesoscale transport equations depending on the mesoscale variables needed to describe a polydisperse multiphase flow. However, it is possible to identify three types that occur in many applications.

1. **Population balance equation (PBE):** In this case, the particles follow the continuous phase so it is not necessary to include their velocity as a mesoscale variable. Letting ξ denote a generic *size* variable, the number density function (NDF) is denoted by $n(t, \mathbf{x}, \xi)$ and represents the number concentration of particles with size in the range ξ to $\xi + \mathrm{d}\xi$ at time t and position $\mathbf{x}$. The NDF is found by solving a PBE:

$$\frac{\partial n}{\partial t} + \frac{\partial}{\partial x_i}\left[u_i(t, \mathbf{x}, \xi)n\right] + \frac{\partial}{\partial \xi}\left[G(t, \mathbf{x}, \xi)n\right] = \frac{\partial}{\partial x_i}\left(D(t, \mathbf{x}, \xi)\frac{\partial n}{\partial x_i}\right) + \mathbb{S} \tag{1}$$

   where repeated Roman indices imply summation. The particle velocity $\mathbf{u}$, growth rate G, diffusivity D and source term $\mathbb{S}$ (e.g., aggregation, breakage, etc.) are known mesoscale models. In this case, the phase space is one dimensional, but in many applications the number of mesoscale variables must be expanded to include other particle properties such as temperature, concentration and surface area. The mesoscale models depend explicitly on ξ and thus the PBE is more complicated than the corresponding macroscale model that depends only on t and $\mathbf{x}$.

2. **Kinetic equation (KE):** In this case, all particle properties except velocity are the same for every particle. The only relevant mesoscale variable is particle velocity $\mathbf{v}$. The NDF is denoted by $n(t, \mathbf{x}, \mathbf{v})$ and found by solving a KE:

$$\frac{\partial n}{\partial t} + \frac{\partial}{\partial x_i}\left(v_i n\right) + \frac{\partial}{\partial v_i}\left[A_i(t, \mathbf{x}, \mathbf{v})n\right] = \mathbb{C} \tag{2}$$

   with a mesoscale acceleration model $\mathbf{A}$ and collision model $\mathbb{C}$. The classical example for mesoscale acceleration is Stokes drag:

$$\mathbf{A} = \frac{1}{\tau_{\mathrm{D}}} \left(\mathbf{u}_{\mathrm{f}} - \mathbf{v}\right) \tag{3}$$

where τ_{D} is characteristic time scale for fluid drag and $\mathbf{u}_{\mathrm{f}}$ is the local fluid velocity for the continuous phase. A widely used model for collisions is the BGK model:

$$\mathbb{C} = \frac{1}{\tau_{\mathrm{C}}} \left(n_{\mathrm{eq}} - n\right) \tag{4}$$

where τ_{C} is a characteristic collision time scale, and n_{eq} is the *equilibrium* NDF. For the KE, the phase space is usually three dimensional, making the total number of independent variables equal to seven.

3. **Generalized population balance equation (GPBE):** The most general type of mesoscale model includes particle velocity and size. The corresponding NDF is denoted as $n(t, \mathbf{x}, \mathbf{v}, \xi)$ and is found by solving a GPBE:

$$\frac{\partial n}{\partial t} + \frac{\partial}{\partial x_i}\left(v_i n\right) + \frac{\partial}{\partial v_i}\left[A_i(t, \mathbf{x}, \mathbf{v}, \xi)n\right] + \frac{\partial}{\partial \xi}\left[G(t, \mathbf{x}, \mathbf{v}, \xi)n\right] = \mathbb{C} \tag{5}$$

with mesoscale models for acceleration $\mathbf{A}$, growth G and collisions and/or aggregation $\mathbb{C}$. Due to the dependence on size, the mesoscale models can be much more complicated than in the monodisperse KE. In particular, collisions leading to aggregation and/or breakage can lead to complex expressions for $\mathbb{C}$ (Marchisio and Fox, 2013). The usual minimum number of mesoscale variables for a GPBE is four, but it is not uncommon to have many more when describing multiphase flows with heat and mass transfer.

1.4 Example macroscale models

Each type of mesoscale model introduced above will lead to a set of moment transport equations that defines the macroscale model. Here we look at the simplest possible forms for each type.

- With a 1-D phase space and 1-D real space, the integer moments of the NDF are defined by

$$M_k = \int_0^\infty \xi^k n \, \mathrm{d}\xi. \tag{6}$$

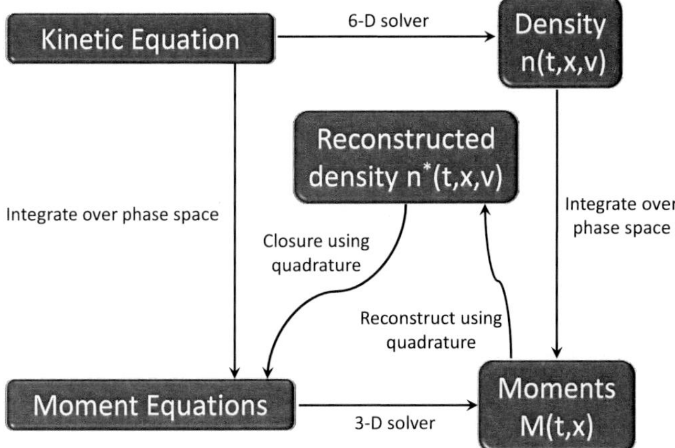

Figure 3. QBMM are used to reconstruct the density function.

Integrating the PBE in (1) over phase space leads to

$$\frac{\partial M_k}{\partial t} + \frac{\partial}{\partial x}\left(\int_0^\infty \xi^k u n \, \mathrm{d}\xi\right) =$$

$$k\int_0^\infty \xi^{k-1} G n \, \mathrm{d}\xi + \frac{\partial}{\partial x}\left(\int_0^\infty \xi^k D \frac{\partial n}{\partial x} \, \mathrm{d}\xi\right) + \int_0^\infty \xi^k \mathbb{S} \, \mathrm{d}\xi \quad (7)$$

where we have assumed that $G \geq 0$ (i.e., particles do not decrease in size). The reader can observe that there are four unclosed terms involving integrals in (7). The first unclosed term (on the left-hand side) is the advective flux of the moments, and will only be closed if u is independent of ξ. Likewise, the diffusive flux involving D will be closed only if the diffusivity does not depend on size. In QBMM, the unclosed terms are closed by reconstructing n from a finite set of moments for each value of t and $\mathbf{x}$. Once the functional form for n is known, the integrals can be evaluated using quadrature formulas. The overall closure procedure is shown in Figure 3.

- With a 1-D phase space and 1-D real space, the integer moments of the NDF are defined by

$$M_k = \int_{-\infty}^{+\infty} v^k n \, \mathrm{d}v. \quad (8)$$

Integrating the KE in (2) over phase space leads to

$$\frac{\partial M_k}{\partial t} + \frac{\partial M_{k+1}}{\partial x} = k \int_{-\infty}^{+\infty} v^{k-1} An \, dv + \int_{-\infty}^{+\infty} v^k \mathbb{C} \, dv \qquad (9)$$

from which we can clearly see that the spatial advection term will never be closed for a fixed maximum value of k. With the simple models in (3) and (4), the integrals on the right-hand side of (9) will be closed, but this need not always be true. With QBMM, the reconstruction shown in Figure 3 is employed to find n from a finite set of moments. Once n is known, M^{k+1} can be found from the definition in (8).

- With a 2-D phase space and 1-D real space, the integer moments of the NDF are defined by

$$M_{kl} = \int_{-\infty}^{+\infty} \int_{0}^{\infty} v^k \xi^l n \, dv d\xi. \qquad (10)$$

Integrating the GPBE in (5) over phase space leads to

$$\frac{\partial M_{kl}}{\partial t} + \frac{\partial M_{k+1l}}{\partial x} = k \int_{-\infty}^{+\infty} \int_{0}^{\infty} v^{k-1} \xi^l An \, dv d\xi$$

$$+ l \int_{-\infty}^{+\infty} \int_{0}^{\infty} v^k \xi^{l-1} Gn \, dv d\xi + \int_{-\infty}^{+\infty} \int_{0}^{\infty} v^k \xi^l \mathbb{C} \, dv d\xi \qquad (11)$$

wherein all of the terms except the first must be closed. With QBMM, it is necessary to reconstruct the joint NDF involving velocity and size. However, once this is done, the reconstructed NDF is used to close the moment transport equation as shown in Figure 3.

1.5 Focus of chapter

The preceding sections provided a brief overview of QBMM and the philosophy behind the closure procedure used to solve the macroscale model. In the remainder of this chapter we will focus on two of the key technical points needed to solve practical problems, namely, how NDF reconstruction is achieved with QBMM and how to solve the closed moment transport equations. While algorithmically different in nature, these two points are strongly coupled by the requirement that the NDF reconstruction can only be successful for a set of realizable moments. In practice, accurate solutions to the moment transport equations require high-order numerical algorithms to reduce numerical diffusion. However, as a general rule, only first-order algorithms will guarantee that numerical errors do not make the moment set unrealizable. It is thus necessary to use only realizable schemes (Marchisio and Fox, 2013), and in Sec. 3 we discuss realizable finite-volume methods. In Sec. 2, we begin by looking at NDF reconstruction algorithms that use a finite set of realizable moments.

2 Quadrature-Based Moment Methods

In this section we discuss moment-inversion algorithms that are used to reconstruct the NDF from a finite set of moments. Some important things to consider when devising an algorithm are as follows.

- Which moments should we choose to solve for?
- What method should we use to reconstruct the NDF?
- How can we extend the algorithm to a multivariate phase space?
- Can we improve the accuracy of the closure without increasing (drastically) the number of moments?

Moreover, since it will be used repeatedly when solving the moment transport equations, we must be able to demonstrate *a priori* that the moment-inversion algorithm is robust and accurate for any arbitrary set of realizable moments.

This section is organized as follows. We start with 1-D phase space where the mathematical theory is well understood and the available algorithms are very robust. Next we look at how these algorithms can be extended to multidimensional phase space where the existing mathematical theory is less well developed. The basic algorithms represent the NDF by a sum of point distributions (i.e., delta functions). Thus, we finish the section by looking at extended algorithms that yield a continuous NDF. The material in this section is covered in much more detail in Marchisio and Fox (2013).

2.1 Gaussian quadrature in 1-D phase space

Let $g(v)$ be a smooth function of v and $n(v) \geq 0$ be an unknown NDF. The formula

$$\int g(v)n(v)\,\mathrm{d}v = \sum_{\alpha=1}^{N} n_\alpha g(v_\alpha) + R_N(g) \tag{12}$$

is a Gaussian quadrature if and only if the N nodes v_α are the roots of an Nth-order orthogonal polynomial $P_N(v)$ ($\perp$ with respect to $n(v)$). The recursion formula for $P_N(v)$ is defined by

$$P_{\alpha+1}(v) = (v - a_\alpha)P_\alpha(v) - b_\alpha P_{\alpha-1}(v), \quad \alpha = 0, 1, 2, \ldots \tag{13}$$

where a_α and b_α are the unique recursion coefficients that define the polynomial. There exists an inversion algorithm for the moments (see Marchisio and Fox (2013) for details)

$$M_k = \int v^k n(v) \mathrm{d}v \tag{14}$$

that proceeds as follows:

$$\{M_0, M_1, \ldots, M_{2N-1}\} \implies \{a_0, a_1, \ldots, a_{N-1}\}, \{b_1, b_2, \ldots, b_{N-1}\}$$
$$\implies \{n_1, n_2, \ldots, n_N\}, \{v_1, v_2, \ldots, v_N\}. \tag{15}$$

In words, given $2N$ sequential integer moments, it is possible to find the recursion coefficients (an ill-conditioned step) and then to find the weights and abscissas (a well-conditioned step). (See Marchisio and Fox (2013) for details.)

Once the weights and abscissas are known, we can use Gaussian quadrature to approximate the unclosed terms in the moment equations. For example,

$$\frac{\mathrm{d}\mathbf{M}}{\mathrm{d}t} = \int \mathbf{S}(v)n(v)\mathrm{d}v \approx \sum_{\alpha=1}^{N} n_\alpha \mathbf{S}(v_\alpha) \tag{16}$$

where $\mathbf{M} = \{M_0, M_1, \ldots, M_{2N-1}\}$ is the moment vector and $\mathbf{S}$ is a generic source term. This Gaussian quadrature approximation has the following properties.

- It is exact if $\mathbf{S}$ is a polynomial of order $\leq 2N - 1$.
- It provides a good approximation for most other cases with a small $N \approx 4$.
- Complications can arise in particular cases (e.g. spatial fluxes, evaporation).
- In all cases, the moments $\mathbf{M}$ must be **realizable** for moment inversion.

The method for solving (16) consists of advancing the moment vector $\mathbf{M}$ in time by evaluating the right-hand side using the weights and abscissas found by inverting the moments at the current time step. For 1-D phase space, the inversion algorithm can be done with low computational cost (see Marchisio and Fox (2013)), making the numerical solution both tractable and accurate.

Note that the Gaussian quadrature is equivalent to a reconstructed N-point distribution function:

$$n^*(v) = \sum_{\alpha=1}^{N} n_\alpha \delta(v - v_\alpha), \tag{17}$$

and the moments will be realizable if $n_\alpha \geq 0$ for all α. As mentioned earlier, with standard Gaussian quadrature the reconstructed NDF is not a smooth function of v, and thus it cannot be evaluated point-wise in phase space. Below, we will introduce an extended quadrature that can be used to reconstruct a smooth NDF.

2.2 Quadrature in multiple dimensions

Unfortunately, there is no method equivalent to Gaussian quadrature for multiple dimensions. However, it is still possible to develop moment-inversion algorithms with the following desirable properties.

- Given a particular moment set $\mathbf{M} = \{M_{ijk} : i, j, k \in 0, 1, \ldots\}$, find non-negative weights n_α and vector abscissas $\mathbf{v}_\alpha$ such that

$$M_{ijk} = \int v_1^i v_2^j v_3^k n(\mathbf{v}) \mathrm{d}\mathbf{v} = \sum_{\alpha=1}^{N} n_\alpha v_{1\alpha}^i v_{2\alpha}^j v_{3\alpha}^k \qquad (18)$$

  for all of the moments in the set.
- If the moment set $\mathbf{M}$ corresponds to an N-point distribution, then the inversion method should be exact (i.e., it should find the exact distribution).
- We should avoid brute-force nonlinear iterative solvers (poor convergence, ill-conditioned, too slow, etc.) in favor of direct solvers (like the 1-D Gaussian quadrature described above).
- The algorithm must yield a realizable quadrature (i.e. non-negative weights with abscissas in the region of phase space where the NDF is nonzero).

The strategy that we have developed for multidimensional quadrature is to choose an optimal moment set to avoid ill-conditioned systems (Fox, 2009b). In the next few sections, we show some example multivariate moment-inversion algorithms that have some or all of the desirable properties.

Brute-force QMOM In the brute-force (BF) quadrature method of moments (QMOM), a nonlinear equation solver is used to solve (18) for an optimal moment set (see Marchisio and Fox (2013) for the definition of optimal moments). For a 2-D phase space, the BFQMOM works as follows. Given $3n^2$ bivariate optimal moments $(n = 2)$:

$$\begin{pmatrix} M_{00} & M_{01} & M_{02} & M_{03} \\ M_{10} & M_{11} & M_{12} & M_{13} \\ M_{20} & M_{21} \\ M_{30} & M_{31} \end{pmatrix}, \qquad (19)$$

solve twelve moment equations:

$$\sum_{\alpha=1}^{4} |n_\alpha| u_\alpha^i v_\alpha^j = M_{ij}$$

to find $\{n_1, \ldots, n_4; u_1, \ldots, u_4; v_1, \ldots, v_4\}$. Because BFQMOM is not direct, the iterative solver often converges slowly (or not at all). Moreover, the system of equations can be singular for (nearly) degenerate cases (e.g., for N-point distributions). In general, BFQMOM is not a good choice for use with QBMM.

Tensor-product QMOM A multivariate tensor-product (TP) QMOM uses multiple 1-D direct quadratures to construct a scaffold in phase space using the TP of the abscissas in each direction (Fox, 2008, 2009a). For a 2-D phase space, TPQMOM proceeds as follows. Given univariate moments, compute 1-D Gaussian quadratures ($n = 2$):

1. $\{M_{00}, M_{10}, M_{20}, M_{30}\} \Longrightarrow \{u_i : i = 1, 2\}$,
2. $\{M_{00}, M_{01}, M_{02}, M_{03}\} \Longrightarrow \{v_j : j = 1, 2\}$.

Using these results, construct $N = 4$ TP abscissas: $\mathbf{v}_\alpha = \{u_i, v_j\} : \alpha = 1, \ldots, 4$. Given the TP abscissas, use $N = 4$ moments $\{M_{ij} : i, j \in 0, 1\}$ to define a linear system for the weights:

$$\mathbf{A}(\mathbf{v}) \begin{bmatrix} n_1 \\ n_2 \\ n_3 \\ n_4 \end{bmatrix} = \begin{bmatrix} M_{00} \\ M_{10} \\ M_{01} \\ M_{11} \end{bmatrix}. \tag{20}$$

The matrix $\mathbf{A}$ will be full rank if the abscissas are distinct, and thus the weights are well defined.

Despite its apparent simplicity, the TPQMOM has some important drawback. First, the solution to the linear system in (20) can yield negative weights. In such cases, it is necessary to develop a correction algorithm to remove the negative weights while solving for the maximum possible number of moments. Another drawback of TPQMOM is that it cannot find the exact N-point distribution in cases where it exists (unless the abscissas just happen to be a tensor product). Finally, even if the weights are positive, TPQMOM in the above example only satisfies eight of the twelve optimal moments in (19). Fortunately it is possible to do better than BF and TPQMOM as we describe next.

Conditional QMOM Currently the best choice for constructing a multidimensional quadrature using the optimal moment set is the conditional quadrature method of moments (CQMOM) (Yuan and Fox, 2011). For a 2-D phase space, the CQMOM works as follows. Define the conditional density function and conditional moments:

$$n(u,v) = f(v|u)n(u) \quad \Longrightarrow \quad \langle V^j | U = u \rangle = \int v^j f(v|u)\,\mathrm{d}v. \tag{21}$$

The strategy behind CQMOM is to find the conditional moments for particular values of u, and then to use the 1-D quadrature method to find the weights and abscissas for each value of u. Thus, we start with a 1-D adaptive quadrature for the U direction ($n = 2$):

$$\{M_{00}, M_{10}, M_{20}, M_{30}\} \Longrightarrow \text{find weights } \rho_\alpha, \text{ abscissas } u_\alpha \tag{22}$$

where the "adaptive" part of the algorithm controls for small weights (Marchisio and Fox, 2013) and reduces the number of abscissas if necessary. For example, if the U direction is a 1-point distribution, only one abscissa is needed and the adaptive QMOM will find the correct abscissa. As noted earlier, the 1-D quadrature representation of $n(u)$ is

$$n(u) = \sum_{\alpha=1}^{2} \rho_\alpha \delta(u - u_\alpha), \tag{23}$$

so that the reconstructed NDF is

$$n(u,v) = \sum_{\alpha=1}^{2} \rho_\alpha \delta(u - u_\alpha) f(v|u_\alpha). \tag{24}$$

It now remains to find a reconstruction for $f(v|u_\alpha)$ for each α.

Using the reconstruction of $n(u,v)$ in (24), the moments are

$$M_{ij} = \int\int \sum_{\alpha=1}^{2} \rho_\alpha u^i v^j \delta(u - u_\alpha) f(v|u_\alpha)\mathrm{d}u\mathrm{d}v = \sum_{\alpha=1}^{2} \rho_\alpha u_\alpha^i \int v^j f(v|u_\alpha)\mathrm{d}v. \tag{25}$$

Thus, once the weights and abscissas in the first direction are known, we can solve linear systems for the conditional moments $\langle V^j | u_\alpha \rangle$:

$$\begin{bmatrix} \rho_1 & \rho_2 \\ \rho_1 u_1 & \rho_2 u_2 \end{bmatrix} \begin{bmatrix} \langle V|u_1 \rangle & \langle V^2|u_1 \rangle & \langle V^3|u_1 \rangle \\ \langle V|u_2 \rangle & \langle V^2|u_2 \rangle & \langle V^3|u_2 \rangle \end{bmatrix} = \begin{bmatrix} M_{01} & M_{02} & M_{03} \\ M_{11} & M_{12} & M_{13} \end{bmatrix}. \tag{26}$$

Thus, in principle, CQMOM controls ten of the twelve optimal moments:

$$\begin{pmatrix} M_{00} & M_{01} & M_{02} & M_{03} \\ M_{10} & M_{11} & M_{12} & M_{13} \\ M_{20} & & & \\ M_{30} & & & \end{pmatrix}, \tag{27}$$

unless the adaptive QMOM reduces the number of abscissas in the first direction. The latter may happen when u_1 and u_2 are nearly equal. The linear system in (27) is well defined as long as $u_1 \neq u_2$, but is ill-conditioned when u_1 and u_2 are nearly equal.

The solution to (26) provides the conditional moments. Then, using 1-D adaptive quadrature in the V direction for each α:

$$\{1, \langle V|u_\alpha\rangle, \langle V^2|u_\alpha\rangle, \langle V^3|u_\alpha\rangle\} \implies \text{find weights } \rho_{\alpha\beta}, \text{ abscissas } v_{\alpha\beta}, \quad (28)$$

we find the final reconstruction for the NDF. The adaptive quadrature may set some of the weights $\rho_{\alpha\beta}$ equal to zero if a subset of conditional moments is not realizable. The reconstructed density is

$$n^*(u,v) = \sum_{\alpha=1}^{2} \sum_{\beta=1}^{2} \rho_\alpha \rho_{\alpha\beta} \delta(u - u_\alpha) \delta(v - v_{\alpha\beta}) \quad (29)$$

and the weights $n_{\alpha\beta} = \rho_\alpha \rho_{\alpha\beta}$ are always non-negative. The extension of the above formulas to include more nodes in each direction is straightforward.

In the formulas given above, we have conditioned on the U direction. Conditioning on $V = v_\alpha$ also uses ten of the twelve optimal moments:

$$\begin{pmatrix} M_{00} & M_{01} & M_{02} & M_{03} \\ M_{10} & M_{11} & & \\ M_{20} & M_{21} & & \\ M_{30} & M_{31} & & \end{pmatrix}. \quad (30)$$

However, the union of two sets is exactly equal to the 2-D optimal moment set. Extension of CQMOM to higher-dimensional phase space is straightforward (Marchisio and Fox, 2013) and also uses the optimal moment set for all permutations of the conditioning variables. For example with $N = 4$ nodes in 2-D, the optimal moment set is

$$\begin{pmatrix} M_{00} & M_{10} & M_{20} & M_{30} \\ M_{01} & M_{11} & M_{21} & M_{31} \\ M_{02} & M_{12} & & \\ M_{03} & M_{13} & & \end{pmatrix}$$

while for $N = 9$ nodes in 2-D the 27 optimal moments are

$$\begin{pmatrix} M_{00} & M_{10} & M_{20} & M_{30} & M_{40} & M_{50} \\ M_{01} & M_{11} & M_{21} & M_{31} & M_{41} & M_{51} \\ M_{02} & M_{12} & M_{22} & M_{32} & M_{42} & M_{52} \\ M_{03} & M_{13} & M_{23} & & & \\ M_{04} & M_{14} & M_{24} & & & \\ M_{05} & M_{15} & M_{25} & & & \end{pmatrix}.$$

In summary, CQMOM offers a direct algorithm for reconstructing a multivariate NDF that is always realizable.

CQMOM with N-point distributions   In addition to being realizable, CQMOM has the additional property that it can reproduce an exact N-point distribution (assuming that at least N abscissas are used). By definition, the 2-D N-point density function (u_α distinct) is

$$n(u,v) = \sum_{\alpha=1}^{N} n_\alpha \delta(u - u_\alpha)\delta(v - v_\alpha) \quad \Longrightarrow \quad f(v|U = u_\alpha) = \delta(v - v_\alpha) \tag{31}$$

where the conditional distribution follows directly from the definition in (21). The moments and conditional moments are thus equal to

$$M_{i0} = \int u^i n(u,v)\mathrm{d}u\mathrm{d}v = \sum_{\alpha=1}^{N} n_\alpha u_\alpha^i \quad \Longrightarrow \quad \langle V|U = u_\alpha \rangle = v_\alpha. \tag{32}$$

Using the 1-D adaptive quadrature for the U direction, we can find the N nodes:

$$M_{i0}, \quad i \in \{0, 1, \ldots, 2N - 1\} \Longrightarrow \text{find weights } n_\alpha, \text{ abscissas } u_\alpha.$$

Here we assume that the u_α are distinct. If this is not the case, the adaptive quadrature will return less than N nodes, in which case the conditional moments in the second direction will yield the missing nodes. The final step is to solve the following linear system for v_α:

$$\begin{bmatrix} n_1 & \cdots & n_N \\ \vdots & & \vdots \\ n_1 u_1^{N-1} & \cdots & n_N u_N^{N-1} \end{bmatrix} \begin{bmatrix} v_1 \\ \vdots \\ v_N \end{bmatrix} = \begin{bmatrix} M_{01} \\ \vdots \\ M_{N-1\,1} \end{bmatrix} \tag{33}$$

where we have used the fact that the first-order conditional moment is just the abscissa. In general, this correspondence between the first-order moment in the second direction and the conditional first-order moment can be used advantageously with CQMOM to solve problems where only the conditional first-order moments are required (e.g., the conditional moment closure used in turbulent combustion).

In Figure 4, examples of 2-D quadratures constructed with the methods described above are presented (Marchisio and Fox, 2013). The exact NDF is a bivariate Gaussian centered at (10,20). When the correlation coefficient ρ is zero, all methods yield the same reconstructed NDF. However, when

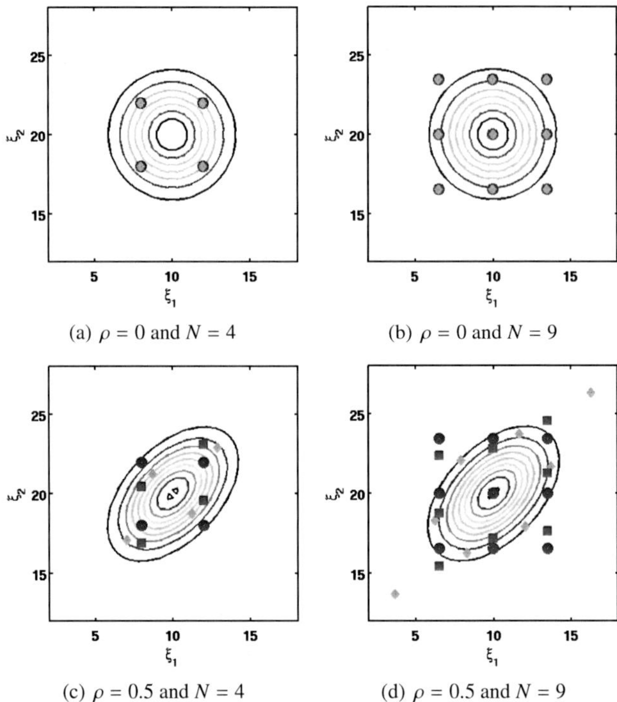

(a) $\rho = 0$ and $N = 4$

(b) $\rho = 0$ and $N = 9$

(c) $\rho = 0.5$ and $N = 4$

(d) $\rho = 0.5$ and $N = 9$

Figure 4. Approximations for bivariate Gaussian: BFQMOM (green diamond) TPQMOM (blue circle) CQMOM (red square).

$\rho \neq 0$, the reconstructed NDFs are different for each method. For this example, when $\rho = 0.5$ the TPQMOM has negative weights for the two points with the smallest probability. Note also that the BFQMOM has two points with very small weights located at large distances from the center. When used with QBMM, such points can result in small time steps for numerical stability, even though they contain very little information about the NDF. Overall, the weights and abscissas found with CQMOM offer the best reconstruction for QBMM.

2.3   Extended quadrature method of moments

The NDF reconstructions presented in the preceding sections use a point distribution, and thus they are not continuous functions. We can improve the reconstructed distribution (Yuan et al., 2012) using kernel density functions (KDF) by defining

$$n(v) = \sum_{\alpha=1}^{N} n_\alpha \delta_\sigma(v, v_\alpha) \tag{34}$$

with N weights $n_\alpha \geq 0$, N abscissas v_α but **only one** spread parameter $\sigma \geq 0$. For example, the KDF can take one of the following forms.

- Gaussian ($-\infty < v < +\infty$) (Chalons et al., 2010):

$$\delta_\sigma(v, v_\alpha) \equiv \frac{1}{\sqrt{2\pi\sigma}} \exp\left[-\frac{(v - v_\alpha)^2}{2\sigma^2}\right]. \tag{35}$$

- Gamma ($0 < v < \infty$) (Yuan et al., 2012):

$$\delta_\sigma(v, v_\alpha) \equiv \frac{v^{\lambda_\alpha-1}e^{-v/\sigma}}{\Gamma(\lambda_\alpha)\sigma^{\lambda_\alpha}} \quad \text{with } \lambda_\alpha = \frac{v_\alpha}{\sigma}. \tag{36}$$

- Beta ($0 < v < 1$) (Yuan et al., 2012):

$$\delta_\sigma(v, v_\alpha) \equiv \frac{v^{\lambda_\alpha-1}(1 - v)^{\mu_\alpha-1}}{B(\lambda_\alpha, \mu_\alpha)} \quad \text{with } \lambda_\alpha = \frac{v_\alpha}{\sigma} \text{ and } \mu_\alpha = \frac{(1 - v_\alpha)}{\sigma}. \tag{37}$$

In the limit $\sigma \to 0$, the KDF behaves like a delta function. The choice to use a single σ is made in order to have a (nearly) direct inversion algorithm whose computational cost is comparable to 1-D Gaussian quadrature. Examples of a two-node beta-EQMOM for different values of σ are shown in Figure 5.

EQMOM algorithm The total number of parameters appearing in N-node EQMOM is $2N + 1$. Thus, an equal number of the moments of $n(v)$ (denoted by m_i) can be controlled. For two-node beta-EQMOM, these moments are

$$m_0 = m_0^*$$

$$m_1 = m_1^*$$

$$m_2 = \frac{1}{1 + \sigma}\left(m_2^* + \sigma m_1^*\right)$$

$$m_3 = \frac{1}{(1 + 2\sigma)(1 + \sigma)}\left(m_3^* + 3\sigma m_2^* + 2\sigma^2 m_1^*\right)$$

$$m_4 = \frac{1}{(1 + 3\sigma)(1 + 2\sigma)(1 + \sigma)}\left(m_4^* + 6\sigma m_3^* + 11\sigma^2 m_2^* + 6\sigma^3 m_1^*\right) \equiv m_{2N}^\dagger(\sigma) \tag{38}$$

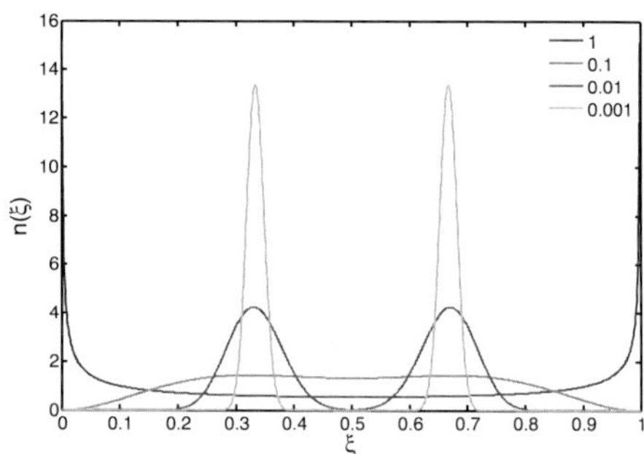

Figure 5. Example beta-EQMOM with $N = 2$ and $n_1 = n_2 = 1/2$, $\xi_1 = 1/3$, $\xi_2 = 2/3$ for different values of σ.

where the starred moments (i.e. the QMOM moments) are defined by

$$m_i^* \equiv \sum_{\alpha=1}^{N} n_\alpha v_\alpha^i. \tag{39}$$

The moments and starred moments can be written as linear systems:

$$m = A(\sigma)m^*$$
$$m^* = A(\sigma)^{-1}m$$

with a coefficient matrix that depends only on σ. Thus, given m_i for $i = 0, \ldots, 2N$, the EQMOM algorithm proceeds as follows (Yuan et al., 2012).

1. Guess σ.
2. Solve for m_i^* for $i = 0, \ldots, 2N - 1$.
3. Solve for n_α and v_α using 1-D Gaussian quadrature with m_i^* for $i = 0, \ldots, 2N - 1$.
4. Compute m_{2N}^* and the resulting estimate $m_{2N}^\dagger$.
5. Iterate on σ until $m_{2N} = m_{2N}^\dagger$.

As described elsewhere (Yuan et al., 2012), a realizability constraint is applied when iterating to find σ with beta- and gamma-EQMOM. Using this algorithm, the first $2N$ moments are always exact with $\max \sigma : m_{2N} \geq m_{2N}^\dagger(\sigma)$. Since the solution to a 1-D nonlinear equation can be done rapidly, the EQMOM moment-inversion algorithm is nearly as fast as 1-D Gaussian quadrature with the advantage that the reconstructed NDF is smooth. Like Gaussian quadrature, EQMOM becomes increasingly more accurate as N is increased.

Closure with EQMOM   With EQMOM, the following unclosed integrals (given n_i, v_i and σ) must be evaluated:

$$\int g(v)n(v)\mathrm{d}v = \sum_{\alpha=1}^{N} n_\alpha \int g(v)\delta_\sigma(v, v_\alpha)\mathrm{d}v \tag{40}$$

where $g(v)$ is an arbitrary smooth function of v (see (12)). Because of the choice for the KDF, we can use classical Gaussian quadrature with known weights $w_{\alpha\beta}$ and abscissas $v_{\alpha\beta}$ to evaluate these integrals:

$$\int g(v)\delta_\sigma(v, v_\alpha)\mathrm{d}v = \sum_{\beta=1}^{N_\alpha} w_{\alpha\beta}g(v_{\alpha\beta}) \tag{41}$$

where N_α can be chosen arbitrarily large to control the integration error (Marchisio and Fox, 2013). The **dual-quadrature representation** of EQMOM is thus

$$n(v) = \sum_{\alpha=1}^{N}\sum_{\beta=1}^{N_\alpha} n_\alpha w_{\alpha\beta}\delta\left(v - v_{\alpha\beta}\right) \quad (N_\alpha = 1 \text{ when } \sigma = 0), \tag{42}$$

and will be exact for integration of polynomials of order $\leq 2N$. We shall see that the dual-quadrature representation has many advantages when deriving numerical schemes for solving the moment transport equations. An important point to keep in mind is that N_α is independent of the number of moments transported so that the additional accuracy attained by using a smooth NDF does not require us to increase greatly the number of transported moments.

Phase-space flux with EQMOM   The need for a continuous NDF is most strongly evident for problems involving a continuous flux in phase space. For example, droplet evaporation is advection in surface-area phase space:

$$\partial_t n + \partial_\xi[R(\xi)n] = 0. \tag{43}$$

The corresponding moments:

$$m_i = \int_0^\infty \xi^i n \, \mathrm{d}\xi \tag{44}$$

obey an unclosed moment equation:

$$\partial_t m_i = -R(0)n(t,0)\delta_{i,0} + i \int_0^\infty \xi^{i-1} R n \, \mathrm{d}\xi \tag{45}$$

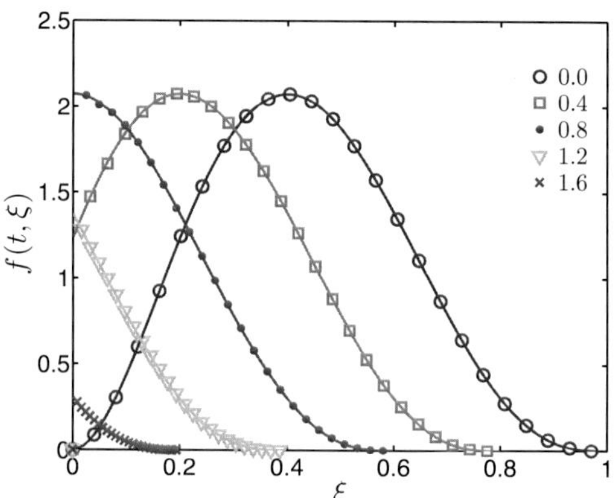

Figure 6. Beta-EQMOM reconstruction at different times starting with a smooth initial NDF: $N = 4$, $N_{1,2} = 80$, $N_{3,4} = 5$ (Yuan et al., 2012).

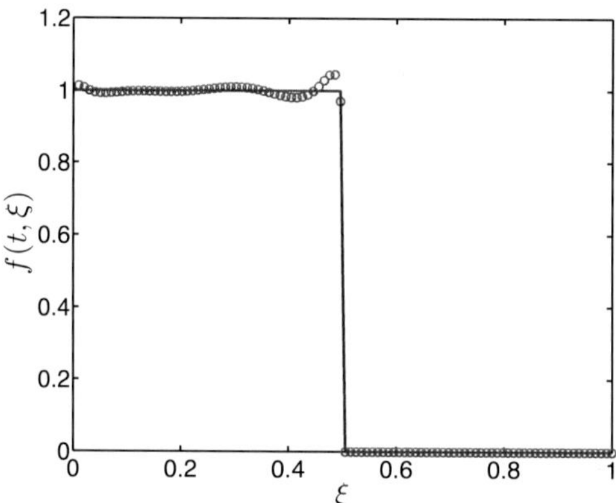

Figure 7. Beta-EQMOM reconstruction at $t = 1/2$ starting with a non-smooth initial NDF: $N = 4$, $N_{1,2} = 80$, $N_{3,4} = 5$ (Yuan et al., 2012).

where $R(0)n(t,0)$ is the rate of loss of particles of zero size. Thus, in order to solve for m_0 we must be able to evaluate the reconstructed NDF at $\xi = 0$, which is not possible with QMOM. However, using the EQMOM dual-quadrature representation, we have

$$n(\xi) = \sum_{\alpha=1}^{N} n_\alpha \delta_\sigma(\xi,\xi_\alpha) \quad \text{and} \quad \delta_\sigma(\xi,\xi_\alpha) \approx \sum_{\beta=1}^{N_\alpha} w_{\alpha\beta}\delta(\xi - \xi_{\alpha\beta}). \tag{46}$$

With pure advection, the weights and abscissas evolve according to

$$\frac{\mathrm{d}w_{\alpha\beta}}{\mathrm{d}t} = 0 \quad \text{and} \quad \frac{\mathrm{d}\xi_{\alpha\beta}}{\mathrm{d}t} = -R(\xi_{\alpha\beta}). \tag{47}$$

The unclosed flux at $\xi = 0$ is found from the "particle" representation as described in Yuan et al. (2012). Example results with $R = 1$ are shown in Figures 6 and 7. By choosing N_α sufficiently large in the dual-quadrature representation, EQMOM predicts the correct evaporative flux at $\xi = 0$ with very little additional computational cost relative to QMOM.

2.4 EQMOM in multiple dimensions

The extension of EQMOM to multiple dimensions makes use of the same ideas as CQMOM (Chalons et al., 2011; Laurant et al., 2013; Marchisio and Fox, 2013; Vié et al., 2012). For example, the bivariate NDF can be written as

$$n(u,v) = \sum_{\alpha=1}^{N} n_\alpha \delta_{\sigma_u}(u,u_\alpha) \left[\sum_{\beta=1}^{N_\alpha} n_{\alpha\beta}\delta_{\sigma_{v,\alpha}}(v,v_{\alpha\beta}) \right] = n(u)f(v|u) \tag{48}$$

with N abscissas u_α, $\mathcal{N} = \sum_{\alpha=1}^{N} N_\alpha$ weights $w_{\alpha\beta} = n_\alpha n_{\alpha\beta} \geq 0$ and $\mathcal{N}$ abscissas $v_{\alpha\beta}$, but **only one** parameter $\sigma_u \geq 0$ and N parameters $\sigma_{v,\alpha}$. The bivariate moments are then given by

$$M_{ij} = \int u^i v^j n(u,v)\mathrm{d}u\mathrm{d}v = \sum_{\alpha=1}^{N}\sum_{\beta=1}^{N_\alpha} w_{\alpha\beta}m_{1,i}^{(\alpha)}m_{2,j}^{(\alpha\beta)} \tag{49}$$

where

$$m_{1,i}^{(\alpha)} \equiv \int u^i \delta_{\sigma_u}(u,u_\alpha)\mathrm{d}u \quad \text{and} \quad m_{2,j}^{(\alpha\beta)} \equiv \int v^j \delta_{\sigma_{v,\alpha}}(v,v_{\alpha\beta})\mathrm{d}v \tag{50}$$

are known functions of the EQMOM parameters found by integrating over the KDF.

Bivariate EQMOM algorithm The parameters appearing in the bivariate EQMOM reconstruction are found using the following algorithm.

1. Apply univariate EQMOM for the moments in u:

$$M_{i0} = \sum_{\alpha=1}^{N} n_\alpha m_{1,i}^{(\alpha)} \quad \text{for } i = 0, \ldots, 2N \implies n_\alpha, u_\alpha \text{ and } \sigma_u. \tag{51}$$

2. Use CQMOM to find the conditional moments defined by

$$\langle V^j \rangle_\alpha \equiv \sum_{\beta=1}^{N_\alpha} n_{\alpha\beta} m_{2,j}^{(\alpha\beta)}. \tag{52}$$

This is done as follows. Starting from the bivariate moments, solve the linear system defined by

$$\sum_{\alpha=1}^{N} n_\alpha m_{1,i}^{(\alpha)} \langle V^j \rangle_\alpha = M_{ij} \quad \text{for } i = 0, \ldots, N-1, \tag{53}$$

for $\langle V^j \rangle_\alpha$ with $j = 1, \ldots, 2N_\alpha$.

3. For each α, apply univariate EQMOM to the conditional moments defined by (52):

$$\{1, \langle V \rangle_\alpha, \ldots, \langle V^{2N_\alpha} \rangle_\alpha\} \implies n_{\alpha\beta}, v_{\alpha\beta} \text{ and } \sigma_{v,\alpha}.$$

This algorithm uses the extended optimal moment set (Marchisio and Fox, 2013). For example, with $\mathcal{N} = 4$ nodes in 2-D phase space, 16 moments are required:

$$\begin{pmatrix} M_{00} & M_{10} & M_{20} & M_{30} & M_{40} \\ M_{01} & M_{11} & M_{21} & M_{31} & M_{41} \\ M_{02} & M_{12} & & & \\ M_{03} & M_{13} & & & \\ M_{04} & M_{14} & & & \end{pmatrix},$$

and with $\mathcal{N} = 9$ nodes in 2-D phase space, 33 moments are required:

$$\begin{pmatrix} M_{00} & M_{10} & M_{20} & M_{30} & M_{40} & M_{50} & M_{60} \\ M_{01} & M_{11} & M_{21} & M_{31} & M_{41} & M_{51} & M_{61} \\ M_{02} & M_{12} & M_{22} & M_{32} & M_{42} & M_{52} & M_{62} \\ M_{03} & M_{13} & M_{23} & & & & \\ M_{04} & M_{14} & M_{24} & & & & \\ M_{05} & M_{15} & M_{25} & & & & \\ M_{06} & M_{16} & M_{26} & & & & \end{pmatrix}.$$

Thus, when applying EQMOM to multivariate problems, only the extended optimal moment set is transported.

Monokinetic EQMOM The definition of multivariate EQMOM in the preceding section treats all variables equally. In some applications, such as particles with very small Stokes number, the velocity will be a unique function of particle size. For such cases, the joint velocity-size NDF can be written as

$$n(v, \xi) = \sum_{\alpha=1}^{N} n_\alpha \delta(v - u(\xi)) \delta_\sigma(\xi, \xi_\alpha) \tag{54}$$

where the size-conditioned velocity $u(\xi)$ is found from velocity-size moments (Marchisio and Fox, 2013). For example, we can approximate the velocity as $u(\xi) = u_0 + u_1\xi + u_2\xi^2 + u_3\xi^3$ with unknowns (u_0, u_1, u_2, u_3). The bivariate moments of the NDF are then equal to

$$M_{i1} = \int \xi^i u(\xi) n \, \mathrm{d}v \mathrm{d}\xi = u_0 M_{i,0} + u_1 M_{i+1,0} + u_2 M_{i+2,0} + u_3 M_{i+3,0}. \tag{55}$$

The velocity parameters can be found by solving a linear system (Marchisio and Fox, 2013). For example, using EQMOM for ξ with $N = 3$ nodes, the linear system is

$$\begin{bmatrix} M_{00} & M_{10} & M_{20} & M_{30} \\ M_{10} & M_{20} & M_{30} & M_{40} \\ M_{20} & M_{30} & M_{40} & M_{50} \\ M_{30} & M_{40} & M_{50} & M_{60} \end{bmatrix} \begin{bmatrix} u_0 \\ u_1 \\ u_2 \\ u_3 \end{bmatrix} = \begin{bmatrix} M_{01} \\ M_{11} \\ M_{21} \\ M_{31} \end{bmatrix}. \tag{56}$$

The coefficient matrix involves only univariate moments and will be non-singular if the NDF for ξ is a smooth function (i.e., if $\sigma > 0$ in the EQMOM reconstruction of $n(\xi)$). This example of the monokinetic EQMOM algorithm requires a total of ten bivariate moments:

$$\{M_{00}, M_{10}, M_{20}, M_{30}, M_{40}, M_{50}, M_{60}, M_{01}, M_{11}, M_{21}, M_{31}\}.$$

It is interesting to note that monokinetic EQMOM uses the same moments as CQMOM for the N-point distribution discussed in Sec. 2.2. This is a result of the fact that $u(\xi)$ is just an approximation of the true size-conditioned velocity, which is the same quantity evaluated at the conditional abscissas in CQMOM.

2.5 Summary of quadrature-based moment methods

To summarize our discussion of QBMM, the reader should keep in mind the following important points.

1. In QBMM, an extended optimal moment set is used to reconstruct the NDF.

2. The NDF must be realizable and the moment-inversion algorithm must be robust.

3. Brute-force QMOM is slow and ill-conditioned, and therefore not recommended for QBMM.

4. Tensor-product QMOM can have negative weights and can not reproduce an N-point NDF.

5. Conditional QMOM is always realizable (but may not reproduce all moments).

6. Extended QMOM gives a smooth NDF with relatively low computational cost.

7. The current "best" moment-inversion algorithms for use with QBMM are as follows.

   - For a 1-D phase space, use univariate EQMOM.
   - For a multivariate phase space, use multivariate EQMOM.
   - With small Stokes numbers (e.g., bubbly flow), use a monokinetic EQMOM.

8. In all cases, the dual-quadrature representation provided by EQMOM should be used to close the integral terms in the GPBE.

The use of QBMM as described above is predicated on the ability to solve the moment transport equations in a manner that guarantees realizable moments. Thus, in the following section, we introduce kinetics-based finite-volume methods that can be used for this purpose.

3 Kinetics-Based Finite-Volume Methods

The moment-inversion algorithms introduced in the preceding section require realizable moment sets, which is only guaranteed with standard first-order finite-volume methods (Desjardins et al., 2006, 2008; Vikas et al., 2012b, 2011a). However, first-order methods suffer from excess numerical diffusion and thus we need to use high-order schemes to produce accurate solutions to the moment transport equations. In this section, we give an overview of kinetics-based finite-volume methods that can achieve high-order accuracy while guaranteeing realizable moments (Vikas et al., 2012b, 2011a). Here, we will assume that QBMM can be applied to the moment set to find a quadrature representation of the NDF of the form

$$n^*(v, \xi) = \sum_{\alpha=1}^{\mathcal{N}} n_\alpha \delta(v - v_\alpha) \delta(\xi - \xi_\alpha) \tag{57}$$

where, for clarity, we limit the phase space to at most two dimensions. (See Marchisio and Fox (2013) for a more general discussion.) The reader can note that the dual-quadrature representation used with EQMOM allows us to express the NDF in the form of (57). Furthermore, for clarity, throughout this section we will limit ourselves to 1-D real space with a uniform grid. The extension to 3-D unstructured grids is described in Vikas et al. (2012b, 2011a).

The set of 2-D extended optimal moments, defined by (10), will be denoted by $\mathbf{M}(t, x)$. To simplify the notation, we introduction a matrix function $\mathbf{K}(v, \xi)$ defined such that

$$\mathbf{M} = \int \mathbf{K}(v, \xi) n \, dv d\xi \quad \Rightarrow \quad K_{ij}(v, \xi) = v^i \xi^j \tag{58}$$

where the integral is over all of 2-D phase space. In words, the expected value of $\mathbf{K}$ is equal to the extended optimal moments. Thus, the expected value of $\mathbf{K}$ times the GPBE is eqaul to the moment transport equations.

3.1   Moment transport equations

Given a set of extended optimal moments, we want to solve the moment transport equation found from the GPBE:

$$\frac{\partial M_{ij}}{\partial t} + \frac{\partial M_{i+1j}}{\partial x} = k \int v^{i-1} \xi^j A n \, dv d\xi + l \int v^i \xi^{j-1} G n \, dv d\xi + \int v^i \xi^j \mathbb{C} \, dv d\xi \tag{59}$$

where the right-hand side is closed using QBMM:

$$\frac{\partial M_{ij}}{\partial t} + \frac{\partial M_{i+1j}}{\partial x} = \sum_{\alpha=1}^{\mathcal{N}} n_\alpha \left\{ i v_\alpha^{i-1} \xi_\alpha^j A_\alpha + j v_\alpha^i \xi_\alpha^{j-1} G_\alpha + v_\alpha^i \xi_\alpha^j \mathbb{C}_\alpha \right\}. \tag{60}$$

Here we assume that the functional forms for the models used to describe acceleration, growth and collisions are known in terms of v and ξ. (See Marchisio and Fox (2013) for numerous examples.) Thus, $A_\alpha = A(v_\alpha, \xi_\alpha)$, etc., are known functions of the abscissas.

When developing a numerical method to solve (60), we must consider the following questions.

- How should we discretize the spatial fluxes on the left-hand side?
- How should we update the moments in time?
- How can we ensure that the moments are always realizable while controlling the numerical diffusion?

In this section, we discuss finite-volume methods that can be used to answer these questions in a general way. These methods are based on the NDF reconstruction in (57) and the underlying properties of the GPBE.

3.2   Kinetics-based spatial fluxes

In order to compute the spatial fluxes we can use a kinetic formulation. For example, the advection equation

$$\partial_t M_{00} + \partial_x M_{10} = 0 \tag{61}$$

has the spatial flux M_{10}, which can be decomposed into two contributions:

$$\begin{aligned} M_{10} &= Q_{10}^- + Q_{10}^+ \\ &= \int_{-\infty}^{0} v \left(\int n^*(v,\xi)\mathrm{d}\xi \right) \mathrm{d}v + \int_{0}^{+\infty} v \left(\int n^*(v,\xi)\mathrm{d}\xi \right) \mathrm{d}v. \end{aligned} \tag{62}$$

In words, the integral over negative velocities is the flux in the negative x direction (i.e., downwind), while the integral over positive velocities is the flux in the positive x direction (i.e., upwind). Using the reconstructed NDF n^*, the downwind and upwind flux components are

$$Q_{10}^- = \sum_{\alpha=1}^{\mathcal{N}} n_\alpha v_\alpha I_{(-\infty,0)}(v_\alpha) \quad \text{and} \quad Q_{10}^+ = \sum_{\alpha=1}^{\mathcal{N}} n_\alpha v_\alpha I_{(0,+\infty)}(v_\alpha) \tag{63}$$

where $I_{\mathbb{S}}(x)$ is the indicator function for the interval $\mathbb{S}$. Because they are defined in terms of n^*, we refer to these fluxes as the kinetics-based fluxes. The extension to arbitrary flux moments is straightforward:

$$\begin{aligned} M_{ij} &= Q_{ij}^- + Q_{ij}^+ \\ &= \int_{-\infty}^{0} v^i \xi^j \left(\int n^*(v,\xi)\mathrm{d}\xi \right) \mathrm{d}v + \int_{0}^{+\infty} v^i \xi^j \left(\int n^*(v,\xi)\mathrm{d}\xi \right) \mathrm{d}v \end{aligned} \tag{64}$$

and

$$Q_{ij}^- = \sum_{\alpha=1}^{\mathcal{N}} n_\alpha v_\alpha^i \xi_\alpha^j I_{(-\infty,0)}(v_\alpha) \quad Q_{ij}^+ = \sum_{\alpha=1}^{\mathcal{N}} n_\alpha v_\alpha^i \xi_\alpha^j I_{(0,+\infty)}(v_\alpha). \tag{65}$$

We can now use these definitions of the spatial fluxes to propose a numerical scheme to solve (60).

3.3   Realizable finite-volume methods

Consider the 1-D advection problem defined for the extended moment set $\mathbf{M}$ by

$$\frac{\partial \mathbf{M}}{\partial t} + \frac{\partial \mathbf{F}(\mathbf{M})}{\partial x} = 0 \tag{66}$$

where the flux function is defined by

$$\mathbf{F}(\mathbf{M}) = \int v\mathbf{K}(v,\xi)n^*(v,\xi)\,\mathrm{d}v\mathrm{d}\xi. \tag{67}$$

Note that $\mathbf{F}$ is indeed a nonlinear function of $\mathbf{M}$ through the moment-inversion algorithm used to reconstruct n^*. The finite-volume representation of the moment vector is

$$\mathbf{M}_k^n \equiv \frac{1}{\Delta x}\int_{x_k-\frac{1}{2}\Delta x}^{x_k+\frac{1}{2}\Delta x} \mathbf{M}(t_n,x)\,\mathrm{d}x \tag{68}$$

where x_k is the center of cell k with uniform width Δx. From this definition, we can observe that $\mathbf{M}_k^n$ is the cell-average value of the extended optimal moment set in cell k at time $t = t_n$. If and only if $\mathbf{M}_k^n$ is a realizable moment set will we be able to use QBMM to reconstruct the corresponding cell-average NDF n_k^n.

Finite-volume formula The first-order, time-explicit, finite-volume formula for (66) is

$$\mathbf{M}_k^{n+1} = \mathbf{M}_k^n - \lambda\left[\mathbf{G}\left(\mathbf{M}_{k+\frac{1}{2},l}^n, \mathbf{M}_{k+\frac{1}{2},r}^n\right) - \mathbf{G}\left(\mathbf{M}_{k-\frac{1}{2},l}^n, \mathbf{M}_{k-\frac{1}{2},r}^n\right)\right] \tag{69}$$

where $\lambda = \Delta t/\Delta x$ and the numerical flux function is defined by

$$\mathbf{G}\left(\mathbf{M}_l, \mathbf{M}_r\right) = \int v^+\mathbf{K}(v,\xi)n_l(v,\xi)\,\mathrm{d}v\mathrm{d}\xi + \int v^-\mathbf{K}(v,\xi)n_r(v,\xi)\,\mathrm{d}v\mathrm{d}\xi \tag{70}$$

with $v^+ = \max(0,v)$ and $v^- = \min(0,v)$. In these expressions, the subscript l corresponds to the left side of the interface between cells and the subscript r corresponds to the right side of the interface. Because the moment transport equation is hyperbolic, the values on the left and right of the interface need not be equal. For example, at the interface $x_k + \Delta x/2$ the moments on the left are $\mathbf{M}_{k+\frac{1}{2},l}^n$ and on the right $\mathbf{M}_{k+\frac{1}{2},r}^n$. The former are related to cell k, while the latter to cell $k+1$. A very important technical point in finite-volume methods is the spatial reconstruction of the moments at the interface (which are unknown) given the known cell-average moments $\mathbf{M}_k^n$. In kinetics-based finite-volume methods, the numerical flux function given by (70) depends on the reconstructed NDF on the left (n_l) and right (n_r) of the cell interface. As we shall see below, the choice of how to define these based on n_k^n will determine whether or not the updated moments $\mathbf{M}_k^{n+1}$ are realizable.

Realizability and spatial fluxes Given $\mathbf{M}_k^n$ we want to define $\mathbf{G}\left(\mathbf{M}_l, \mathbf{M}_r\right)$ to achieve high-order spatial accuracy, but at the same time keeping $\mathbf{M}_k^{n+1}$ realizable. We can define the discrete distribution function h_k as

$$\mathbf{M}_k^{n+1} \equiv \int \mathbf{K}(v, \xi) h_k(v, \xi)\, \mathrm{d}v \mathrm{d}\xi. \tag{71}$$

Using this definition on the left-hand side of (69), the finite-volume formula can be rewritten as

$$h_k = \left(\lambda|v^-|n_{k+\frac{1}{2},r}^n + \lambda v^+ n_{k-\frac{1}{2},l}^n + n_k^n\right) - \lambda|v^-|n_{k-\frac{1}{2},r}^n - \lambda v^+ n_{k+\frac{1}{2},l}^n. \tag{72}$$

A sufficient condition for realizable moments $\mathbf{M}_k^{n+1}$ is that $h_k \geq 0$ for all v, ξ and k. In (72), the terms inside the parentheses are always nonnegative. Thus, the final two terms will determine whether h_k is nonnegative.

In a first-order finite-volume scheme, the cell interface values are the same as the cell-average values in the same cell:

$$n_{k-\frac{1}{2},r}^n = n_{k+\frac{1}{2},l}^n = n_k^n \tag{73}$$

so that (72) becomes

$$h_k = \lambda|v^-|n_{k+1}^n + \lambda v^+ n_{k-1}^n + \left(1 - \lambda|v^-| - \lambda v^+\right) n_k^n. \tag{74}$$

Thus, h_k will be nonnegative if the following realizability condition is satisfied:

$$\frac{1}{|v^-| + v^+} \geq \lambda \quad \text{for all } k. \tag{75}$$

This condition is the same as the stability condition on the time step size. With the first-order scheme, the moments are realizable, but the scheme is very diffusive and an extremely fine grid is needed to get reasonably accurate results.

In order to reduce the numerical diffusion, we can use a quasi-high-order scheme based on the quadrature reconstruction of the NDF. Let

$$n_k^n = \sum_\alpha \rho_{\alpha,k}^n \delta(v - v_{\alpha,k}^n)\delta(\xi - \xi_{\alpha,k}^n) \tag{76}$$

where $v_{\alpha,k}^n$ and $\xi_{\alpha,k}^n$ are the abscissas found from the cell-average moments $\mathbf{M}_k^n$ using QBMM. We then define the interface NDF as

$$n_{k-\frac{1}{2},r}^n = \sum_\alpha \rho_{\alpha,k-\frac{1}{2},r}^n \delta(v - v_{\alpha,k}^n)\delta(\xi - \xi_{\alpha,k}^n) \tag{77}$$

and

$$n^n_{k+\frac{1}{2},l} = \sum_\alpha \rho^n_{\alpha,k+\frac{1}{2},l}\delta(v - v^n_{\alpha,k})\delta(\xi - \xi^n_{\alpha,k}), \tag{78}$$

where the abscissas use the first-order formula but the weights use a high-order spatial reconstruction. Substituting in (72), we find

$$h_k = \lambda|v^-|n^n_{k+\frac{1}{2},r} + \lambda v^+ n^n_{k-\frac{1}{2},l}$$
$$+ \sum_\alpha \left(\rho^n_{\alpha,k} - \lambda|v^-|\rho^n_{\alpha,k-\frac{1}{2},r} - \lambda v^+ \rho^n_{\alpha,k+\frac{1}{2},l} \right) \delta(v - v^n_{\alpha,k})\delta(\xi - \xi^n_{\alpha,k}). \tag{79}$$

This reconstruction will be nonnegative if the following realizability condition is satisfied:

$$\min_\alpha \left(\frac{\rho^n_{\alpha,k}}{|v^-_{\alpha,k}|\rho^n_{\alpha,k-\frac{1}{2},r} + v^+_{\alpha,k}\rho^n_{\alpha,k+\frac{1}{2},l}} \right) \geq \lambda. \tag{80}$$

We can use any available high-order, finite-volume scheme for the spatial reconstruction of the interface weights $\rho^n_{\alpha,k-\frac{1}{2},r}$ and $\rho^n_{\alpha,k+\frac{1}{2},l}$. For example, a second-order spatial reconstruction uses

$$\rho^n_{\alpha,k-\frac{1}{2},r} = \rho^n_{\alpha,k} - \frac{1}{2}S^n_{\alpha,k}\Delta x$$
$$\rho^n_{\alpha,k+\frac{1}{2},l} = \rho^n_{\alpha,k} + \frac{1}{2}S^n_{\alpha,k}\Delta x \tag{81}$$

where $S^n_{\alpha,k}$ is the estimate for the gradient of the weight α in the cell k (Marchisio and Fox, 2013). These slope estimates are found using the values in the neighboring cells along with a slope limiter to avoid overshoots.

In summary, the quasi-high-order schemes result in an explict formula for h_k from which we can guarantee realizability by applying a realizability condition on the time step (which is usually more restrictive that the stability condition). In practice, the time step Δt at step n is fixed as a fraction (e.g., 1/2) of the first-order condition in (75):

$$\Delta t = \frac{\Delta x}{2\max_{\alpha,k}|v^n_{\alpha,k}|}, \tag{82}$$

which fixes the value of $\lambda = \Delta x/\Delta t$ at step n. Then, if λ satisfies (80) in cell k, a second-order spatial reconstruction is used in that cell. Otherwise, the first-order spatial reconstruction is used. Note that this procedure is applied at each step n so that the order of the spatial reconstruction used in a given cell may change as the simulation progresses.

3.4 Realizable time-stepping schemes

In the discussion above, we have used the first-order explicit time-stepping scheme:

$$\mathbf{M}_k^{n+1} = \mathbf{M}_k^n - \lambda \left[\mathbf{G}\left(\mathbf{M}_{k+\frac{1}{2},l}^n, \mathbf{M}_{k+\frac{1}{2},r}^n\right) - \mathbf{G}\left(\mathbf{M}_{k-\frac{1}{2},l}^n, \mathbf{M}_{k-\frac{1}{2},r}^n\right) \right], \quad (83)$$

which is always realizable if the moments $\mathbf{M}_k^n$ are computed from the quadrature formula

$$\mathbf{M}_k^n = \sum_\alpha \rho_{\alpha,k}^n \mathbf{K}(v_{\alpha,k}^n, \xi_{\alpha,k}^n). \quad (84)$$

To go to higher order in time, we might try the second-order Runga-Kutta (RK2); however, it does not always give realizable moments (Vikas et al., 2011a). Fortunately, the strong stability preserving version (RK2SSP) defined by

$$\mathbf{M}_k^* = \mathbf{M}_k^n - \lambda \left[\mathbf{G}\left(\mathbf{M}_{k+\frac{1}{2},l}^n, \mathbf{M}_{k+\frac{1}{2},r}^n\right) - \mathbf{G}\left(\mathbf{M}_{k-\frac{1}{2},l}^n, \mathbf{M}_{k-\frac{1}{2},r}^n\right) \right]$$

$$\mathbf{M}_k^{**} = \mathbf{M}_k^* - \lambda \left[\mathbf{G}\left(\mathbf{M}_{k+\frac{1}{2},l}^*, \mathbf{M}_{k+\frac{1}{2},r}^*\right) - \mathbf{G}\left(\mathbf{M}_{k-\frac{1}{2},l}^*, \mathbf{M}_{k-\frac{1}{2},r}^*\right) \right] \quad (85)$$

$$\mathbf{M}_k^{n+1} = \frac{1}{2}\left(\mathbf{M}_k^n + \mathbf{M}_k^{**}\right)$$

is always realizable (Vikas et al., 2011a). This is because the first two steps are exactly the same as the first-order formula, and the third step is just the average of two moment sets. Since moment space is convex, all three steps are guaranteed to be realizable if λ is chosen to satisfy the realizability condition. In summary, by generalizing the methods described above it is possible to achieve second-order in space and time on unstructured grids (Vikas et al., 2012b, 2011a).

3.5 Summary of kinetics-based finite-volume methods

We end this section with a review of the most important points.

1. When solving moment transport equations, we must guarantee realizability in order to have robust solutions.
2. First-order finite-volume methods are realizable, but too diffusive for most applications.
3. Standard high-order finite-volume methods lead to unrealizable moments.
4. Kinetics-based flux functions can be designed to be realizable for arbitrary moment sets.

5. The dual-quadrature representation with high-order spatial reconstruction of the weights is recommended for solving the moment transport equations.

6. High-order time-stepping schemes are also possible, but must be checked for realizability.

7. Kinetics-based finite-volume methods provide a robust treatment of shocks and discontinuous solutions on unstructured grids.

Having described QBMM and the algorithms used to solve moment transport equations, we will now look at an application of these methods.

4 Application to Gas-Particle Flow

In this section we present a brief overview of the mesoscale models used to describe gas-particle flows. For clarity, we will consider only the simplest forms of the physical processes, namely, fluid drag and gravity for acceleration, and hard-sphere collisions.

4.1 Gas-particle flow model

Consider a disperse phase composed of mono- or bidisperse particles (e.g. different diameters and/or densities). The disperse-phase KE for the ith particle type can be written as

$$\frac{\partial n_i}{\partial t} + \mathbf{v} \cdot \frac{\partial n_i}{\partial \mathbf{x}} + \frac{\partial}{\partial \mathbf{v}} \cdot (\mathbf{A}_i n_i) = \mathbb{C}_{ii} + \mathbb{C}_{ij} \tag{86}$$

where $n_i(t, \mathbf{x}, \mathbf{v})$ is the velocity NDF for particle type $i = 1, 2$, $\mathbf{v}$ is the particle velocity (which is the only mesoscale variable because the particle properties do not change), $\mathbf{A}_i$ is the acceleration model (drag, gravity) for type i particles, and $\mathbb{C}_{ii}$, $\mathbb{C}_{ij}$ are the rates of change of n_i due to particle-particle collisions between like and unlike particles, respectively. By definition, the zero-order moments are the particle-phase volume fractions for each particle type:

$$\alpha_i = \int n_i \mathrm{d}\mathbf{v} \tag{87}$$

and we define the disperse-phase volume fraction as $\alpha_\mathrm{p} = \alpha_1 + \alpha_2$. The first-order moments define the disperse-phase velocities for each particle type $\mathbf{U}_{\mathrm{p}i}$:

$$\alpha_i \mathbf{U}_{\mathrm{p}i} = \int \mathbf{v} n_i \mathrm{d}\mathbf{v}. \tag{88}$$

The total mass and momentum of the disperse phase can be related to the corresponding quantities for the particle types by using the material densities ρ_i:

$$\rho_{\mathrm{p}} = \rho_1 \alpha_1 + \rho_2 \alpha_2, \tag{89}$$

$$\rho_{\mathrm{p}} \mathbf{U}_{\mathrm{p}} = \rho_1 \alpha_1 \mathbf{U}_{\mathrm{p}1} + \rho_2 \alpha_2 \mathbf{U}_{\mathrm{p}1}. \tag{90}$$

In the following, we will assume that there is no mass transfer between particle types or between phases. Thus, only momentum transfer between phases is considered in the fluid mass and momentum balances.

The disperse-phase KE is coupled to the gas-phase governing equations:

$$\frac{\partial}{\partial t} \left(\rho_{\mathrm{g}} \alpha_{\mathrm{g}} \right) + \nabla \cdot \left(\rho_{\mathrm{g}} \alpha_{\mathrm{g}} \mathbf{U}_{\mathrm{g}} \right) = 0, \tag{91}$$

$$\frac{\partial}{\partial t} \left(\rho_{\mathrm{g}} \alpha_{\mathrm{g}} \mathbf{U}_{\mathrm{g}} \right) + \nabla \cdot \left(\rho_{\mathrm{g}} \alpha_{\mathrm{g}} \mathbf{U}_{\mathrm{g}} \mathbf{U}_{\mathrm{g}} \right) = \nabla \cdot \alpha_{\mathrm{g}} \boldsymbol{\tau}_{\mathrm{g}} + \boldsymbol{\beta}_{\mathrm{g}} + \rho_{\mathrm{g}} \alpha_{\mathrm{g}} \mathbf{g}, \tag{92}$$

where the gas density ρ_{g} is assumed constant, the gas-phase volume fraction is $\alpha_{\mathrm{g}} = 1 - \alpha_{\mathrm{p}}$, and the exchange of momentum with the disperse phase due to the fluid drag is $\boldsymbol{\beta}_{\mathrm{g}}$. The conservation of momentum between the gas and particles yields the following definition:

$$\boldsymbol{\beta}_{\mathrm{g}} = -\sum_{i=1}^{2} \rho_i \int (\mathbf{A}_i - \mathbf{g}) n_i \mathrm{d}\mathbf{v} \tag{93}$$

where $\mathbf{A}_i - \mathbf{g}$ is the acceleration model for drag (i.e., without body forces). The fluid-phase models given above are the simplest possible extension of the Navier-Stokes equations for single-phase flow. The fluid stress tensor $\boldsymbol{\tau}_{\mathrm{g}}$ can be modeled as a Newtonian fluid with an effective viscosity that accounts for the pseudo-turbulent velocity fluctuations generated by individual particle wakes (Marchisio and Fox, 2013; , 2010). An important point to note is that the KE is coupled to the gas-phase model only through low-order moments of the NDF. For this reason, a closure of the KE that accurately predicts the low-order moments should yield good predictions for gas-particle flows.

4.2   Collision models

The collision terms in (86) require a mesoscale model to achieve closure in terms of n_i. However, conservation of mass and momentum during collisions yield the following constraints on the model:

$$\int \mathbb{C}_{ii} \mathrm{d}\mathbf{v} = 0, \quad \int \mathbb{C}_{ij} \mathrm{d}\mathbf{v} = 0,$$
$$\int \mathbf{v} \mathbb{C}_{ii} \mathrm{d}\mathbf{v} = 0, \quad \int \mathbf{v} (\rho_1 \mathbb{C}_{12} + \rho_2 \mathbb{C}_{21}) \mathrm{d}\mathbf{v} = 0. \tag{94}$$

The last constraint is a statement of conservation of momentum during collisions of unlike particles. However, during such collisions, the momentum of a given particle type can change, resulting in a net exchange of momentum between particle types. We now look at two mesoscale models for the collision terms.

Boltzmann-Enskog inelastic, hard-sphere collision integral The exact Boltzmann-Enskog collision integral for monodisperse hard-sphere collisions is (Fox and Vedula, 2010; Marchisio and Fox, 2013)

$$\mathbb{C} = \frac{6}{\pi d_{\mathrm{p}}} \int_{\mathbb{R}^3} \int_{\mathbb{S}^+} \left[\chi f^{(2)}(\mathbf{x}, \mathbf{v}_1''; \mathbf{x} - d_{\mathrm{p}}\mathbf{n}, \mathbf{v}_2'') \right.$$

$$\left. - f^{(2)}(\mathbf{x}, \mathbf{v}_1; \mathbf{x} + d_{\mathrm{p}}\mathbf{n}, \mathbf{v}_2) \right] |\mathbf{k} \cdot \mathbf{n}| \, \mathrm{d}\mathbf{n}\mathrm{d}\mathbf{v}_2 \quad (95)$$

where $f^{(2)}$ is the two-particle density function, d_{p} is the particle diameter, $\mathbf{k}$ is the relative velocity vector, $\mathbf{n}$ is the unit vector along the direction of particles centers, $\mathbb{S}^+$ is unit half sphere where $\mathbf{k} \cdot \mathbf{n} > 0$, and χ is a factor relating pre- and post-collisional velocities. There are two important issues related to the closure of this collision integral. First, the two-particle density function is not known, so it must be approximated in terms of the known one-particle NDF. Second, the two-particle density function in (95) is evaluated at two different spatial locations separated by the particle diameter d_{p}. The usual closure expands the two-particle density function around the point of contact (Fox and Vedula, 2010; Marchisio and Fox, 2013), which leads to two terms referred to as point collisions and the collisional flux. The point collision term is the same as the Boltzmann collision integral, which approximates the two-particle density by

$$f^{(2)}(\mathbf{x}, \mathbf{v}_1; \mathbf{x} + d_{\mathrm{p}}\mathbf{n}, \mathbf{v}_2) \approx f^{(2)}(\mathbf{x}, \mathbf{v}_1; \mathbf{x}, \mathbf{v}_2) \approx n(\mathbf{x}, \mathbf{v}_1)n(\mathbf{x}, \mathbf{v}_2). \quad (96)$$

This collision model is valid for very dilute systems where the average distance between particles is very large compared to d_{p}. In contrast, the collisional flux term becomes important for moderately dense systems wherein the finite size of the particles cannot be ignored (Fox and Vedula, 2010). The closure of the Boltzmann-Enskog collision integral for polydisperse particles is described in Marchisio and Fox (2013).

Even with the inelastic Boltzmann model for the collisions:

$$\mathbb{C}(\mathbf{v}_1) = \frac{6}{\pi d_{\mathrm{p}}} \int_{\mathbb{R}^3} \int_{\mathbb{S}^+} \left[\chi \, n(\mathbf{v}_1'')n(\mathbf{v}_2'') - n(\mathbf{v}_1)n(\mathbf{v}_2) \right] |\mathbf{k} \cdot \mathbf{n}| \, \mathrm{d}\mathbf{n}\mathrm{d}\mathbf{v}_2, \quad (97)$$

the remaining integrals over $\mathbf{n}$ and $\mathbf{v}_2$ are difficult to work with in a numerical solver. Using QBMM, the integral over $\mathbf{n}$ can be done analytically for integer moments (Fox and Vedula, 2010; Marchisio and Fox, 2013), and then the moments of $\mathbb{C}(\mathbf{v})$ can be computed using quadrature (Icardi et al., 2012; Marchisio and Fox, 2013; Passalacqua and Fox, 2012; Passalacqua et al., 2011). As shown in Passalacqua and Fox (2012), the full Boltzmann collision integral is needed for polydisperse systems when the particle diameters are different. Otherwise, for monodisperse systems, a simpler linear model can be used.

Linear collision model   For monodisperse particles, the collision term can be approximated by the linear model given in (4). The collision time scale for hard-sphere collisions is

$$\tau_{\mathrm{C}} = \frac{\gamma \sqrt{\pi} d_{\mathrm{p}}}{12 g_0 \alpha_{\mathrm{p}} \Theta_{\mathrm{p}}^{1/2}}, \tag{98}$$

and the inelastic equilibrium distribution (Marchisio and Fox, 2013) is

$$n_{\mathrm{eq}} = \frac{\alpha_{\mathrm{p}}}{\left[\det\left(2\pi\boldsymbol{\lambda}\right)\right]^{1/2}} \exp\left[-\frac{1}{2}(\mathbf{v} - \mathbf{U}_{\mathrm{p}}) \cdot \boldsymbol{\lambda}^{-1} \cdot (\mathbf{v} - \mathbf{U}_{\mathrm{p}})\right]. \tag{99}$$

The covariance matrix $\boldsymbol{\lambda}$ is defined by

$$\boldsymbol{\lambda} = \gamma\omega^2 \Theta_{\mathrm{p}} \mathbf{I} + \left(\gamma\omega^2 - 2\gamma\omega + 1\right) \boldsymbol{\sigma}_{\mathrm{p}} \tag{100}$$

where $\omega = (1 + e)/2$ and e is the coefficient of restitution for particle-particle collisions. For $e = 1$, the parameter γ can take on values in the range $0 < \gamma < 3/2$. The value of $\gamma = 1$ corresponds to the BGK model, and $\gamma = 3/2$ is the ES-BGK model (Passalacqua et al., 2011). The remaining variables in (100) are the velocity covariance matrix $\boldsymbol{\sigma}_{\mathrm{p}}$, the granular temperature Θ_{p} and the identity tensor $\mathbf{I}$. In (98) g_0 is the radial distribution function that depends on the value of α_{p}. In fact, due to the dependence of τ_{C} on α_{p} and Θ_{p}, the collision model is nonlinear in the moments of the NDF. The linear collision model yields good approximations to the hard-sphere collision integral when the velocity distribution is near the equilibrium distribution (i.e., when the collision term is dominant in the kinetic equation). For highly non-equilibrium systems, the full Boltzmann collision integral with quadrature-based closure is needed for good results (Icardi et al., 2012). Unfortunately, for polydisperse systems the linear models do not perform adequately (Passalacqua and Fox, 2012) and thus the full Boltzmann integral is preferred for such systems whenever the term for collisions is dominant.

4.3 Dimensionless numbers and flow regimes

From the KE for the disperse phase and the mass and momentum balance for the gas phase, it is possible to identify the following dimensionless numbers that will be important for determining the flow characteristics.

- The gas-phase Reynolds number determines whether the single-phase gas flow is laminar or turbulent. It is defined as

$$\mathrm{Re_g} = \frac{\rho_g U_g L_g}{\mu_g}$$

  where the gas-phase time scale is written in terms of a characteristic velocity U_g and a characteristic length scale L_g. μ_g is the gas-phase viscosity and ρ_g is the gas density.

- The particle Stokes number determines how fast a particle will respond to changes in the fluid motion. It is defined as

$$\mathrm{St_p} = \frac{\rho_p d_p^2 U_g}{18\mu_g L_g}$$

  where d_p is the particle diameter and ρ_p is the particle material density.

- The particle-phase Mach number, which determines whether the particle transport is convective versus diffusive, is defined by

$$\mathrm{Ma_p} = \frac{|\mathbf{U}_p|}{\Theta_p^{1/2}}$$

  where Θ_p is the granular temperature.

- The particle Knudsen number determines whether the particle flow is collisional or free transport. It is defined by

$$\mathrm{Ma_p} < 1 \Rightarrow \mathrm{Kn_p} = \sqrt{\frac{\pi}{2}} \frac{\tau_C \Theta_p^{1/2}}{L_p}$$

or

$$\mathrm{Ma_p} > 1 \Rightarrow \mathrm{Kn_p} = \sqrt{\frac{\pi}{2}} \frac{\tau_C |\mathbf{U}_p|}{L_p}$$

  where L_p is the characteristic length scale for the particle phase. The hydrodynamic regime corresponds to very small Knudsen number and occurs when collisions occur much faster than all other physical processes.

Gas-particle flows with $\mathrm{Ma_p} \gg 1$ are very common because the granular temperature is often small due to the fluid drag. It is thus necessary to use numerical methods adapted to high-Mach-number flows (such as finite-volume methods).

In the hydrodynamic limit, the Chapman-Enskog expansion can be used to simplify the moment transport equations. Indeed, for very small Knudsen numbers, only the zero- and first-order moments (e.g., conservation of mass and momentum) and the trace of the second-order moments (total energy) are needed to describe the flow. For example, the following two regimes are often cited in the literature on granular flows.

- Continuous regime ($\mathrm{Kn_p} < 0.01$): Navier-Stokes-Fourier (NSF) equations with no-slip BC.
- Slip regime ($0.01 < \mathrm{Kn_p} < 0.1$): NSF equations with partial-slip conditions at walls

For $\mathrm{Kn_p} > 0.1$: Higher-order approximations of the kinetic equation or direct solutions are required. Gas-particle flows with $\mathrm{Kn_p} \gg 0.1$ are very common. For such flows, the above-mentioned regimes do not apply and the full set of moment equations should be solved using QBMM and realizable finite-volume schemes.

Based on the dimensionless numbers introduced above, we can classify gas-particle flows into the following regimes.

- Very dilute (or weakly collisional) flow:
    - $\mathrm{Kn_p} \gg 1$
    - volume fraction $< 1\%$
    - very small Θ_p (e.g., due to small $\mathrm{St_p}$) and $\mathrm{Ma_p} > 1$
- Dilute flow:
    - $0 < \mathrm{Kn_p} < 1$
    - volume fraction $< 1\%$ (negligible collisional flux)
    - moderate Θ_p
- Moderately dense flow:
    - $0 < \mathrm{Kn_p} < 1$
    - volume fraction $> 1\%$ (non-negligible collisional flux due to $g_0 > 1$)
    - moderate Θ_p
- Dense flow:
    - $0 < \mathrm{Kn_p} \ll 1$
    - volume fraction $> 20\%$ (strong collisional flux due to $g_0 \gg 1$)
    - moderate Θ_p

The reader should keep in mind that all of these regimes could occur in the same physical system. For example, the system could contain a fluidized bed (dense flow), a riser (dilute and very dilute), and a cyclone (moderately dense). Thus, in order to simulation the entire system with the same code, it will be necessary to use numerical methods that can handle all regimes simultaneously.

4.4 Lagrangian versus Eulerian simulations

For gas-particle flows, it is possible to use either Lagrangian or Eulerian simulation methods. For example, the kinetic equation

$$\frac{\partial n}{\partial t} + \mathbf{v} \cdot \frac{\partial n}{\partial \mathbf{x}} + \frac{\partial}{\partial \mathbf{v}} \cdot (\mathbf{A}n) = \mathbb{C} \tag{101}$$

can be simulate by a Lagrangian method. For a large ensemble of sample particles, the position and velocity of particle α are tracked using the equations

$$\begin{aligned} \frac{\mathrm{d}\mathbf{x}^{(\alpha)}}{\mathrm{d}t} &= \mathbf{v}^{(\alpha)} \\ \frac{\mathrm{d}\mathbf{v}^{(\alpha)}}{\mathrm{d}t} &= \mathbf{A}^{(\alpha)} + \mathcal{C}^{(\alpha)} \end{aligned} \tag{102}$$

where $\mathcal{C}^{(\alpha)}$ is a Lagrangian collision model that generates the same evolution for the moments as $\mathbb{C}$. The principal limitation of Lagrangian approaches is the statistical noise present due to the finite sample size, and coupling errors caused by passing noisy data back to the Eulerian gas-phase solver. The latter are especially important when the mass loading is high enough to result in significant momentum coupling between phases.

In the Eulerian method, a finite set of velocity moments are tracked:

$$M^0 = \alpha_\mathrm{p} = \int n \, \mathrm{d}\mathbf{v},$$

$$M_i^1 = \alpha_\mathrm{p} U_{\mathrm{p}i} = \int v_i \, n \, \mathrm{d}\mathbf{v},$$

$$\vdots$$

$$M_{ij\ldots}^\gamma = \int (v_i v_j \cdots) \, n \, \mathrm{d}\mathbf{v},$$

and the unknown moments closed using QBMM. While the Eulerian method is free from statistical noise and can handle strongly coupled flows in an accurate and robust manner, the accuracy of the moment closure limits the overall accuracy of the Eulerian simulation. In practice, Lagrangian simulations of canonical flows can be used to test the moment closures employed in Eulerian methods. The latter can then be applied to complex non-canonical flows that occur in practical applications.

For the Eulerian solver, the transport equations for the extended optimal moment set with $\mathcal{N} = 8$ nodes can be obtained from the kinetic equation in (101):

$$\frac{\partial M^0}{\partial t} + \frac{\partial M_i^1}{\partial x_i} = 0$$

$$\frac{\partial M_i^1}{\partial t} + \frac{\partial}{\partial x_j}\left(M_{ij}^2 + G_{ij}^2\right) = g_i M^0 + D_i^1$$

$$\frac{\partial M_{ij}^2}{\partial t} + \frac{\partial}{\partial x_k}\left(M_{ijk}^3 + G_{ijk}^3\right) = g_i M_j^1 + g_j M_i^1 + D_{ij}^2 + C_{ij}^2$$

$$\frac{\partial M_{ijk}^3}{\partial t} + \frac{\partial}{\partial x_l}\left(M_{ijkl}^4 + G_{ijkl}^4\right) = g_i M_{jk}^2 + g_j M_{ik}^2 + g_k M_{ij}^2$$
$$+ D_{ijk}^3 + C_{ijk}^3$$

$$\frac{\partial M_{ijkl}^4}{\partial t} + \frac{\partial}{\partial x_m}\left(M_{ijklm}^5 + G_{ijklm}^5\right) = g_i M_{jkl}^3 + g_j M_{ikl}^3 + g_k M_{ijl}^3 + g_l M_{ijk}^3$$
$$+ D_{ijkl}^4 + C_{ijkl}^4$$

$$(103)$$

where (i, j, k, l) take on integer values corresponding to the optimal moments and repeated indices imply summation. As described earlier, QBMM are used to close the spatial fluxes, drag and collision terms. In (103), the terms on the left-hand side denoted by G are the collisional fluxes, and must be included to treat moderately dense flows (Fox and Vedula, 2010). Note that (103) is valid for monodisperse particles, for which the linear collision model provides good accuracy (Passalacqua and Fox, 2012).

In the very dilute regime, collisions are negligible and hence the hydro-dynamic model is not valid. However, if the hydrodynamic model is used to simulate very dilute jets in a cross flow, the result shown in Figure 8 is found. The hydrodynamic model uses the following five moment equations

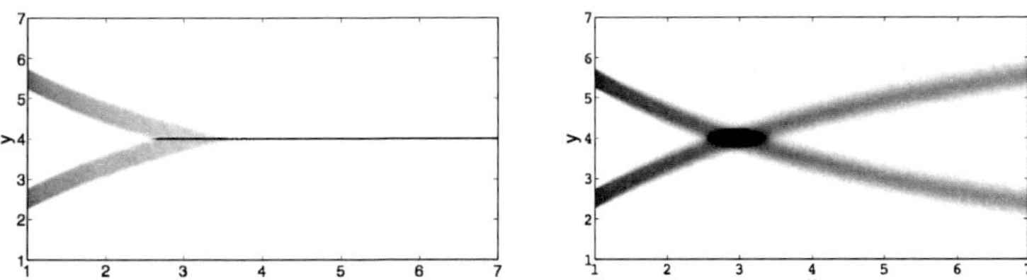

Figure 8. Hydrodynamic model (left) predicts artificial collisions where there are none with the QBMM closure (right).

(Fan and Fox, 2008):

$$\frac{\partial M^0}{\partial t} + \frac{\partial M_i^1}{\partial x_i} = 0,$$

$$\frac{\partial M_i^1}{\partial t} + \frac{\partial}{\partial x_j}\left(M_{ij}^2 + G_{ij}^2\right) = g_i M^0 + D_i^1, \qquad (104)$$

$$\frac{\partial M_{ii}^2}{\partial t} + \frac{\partial}{\partial x_j}\left(M_{iij}^3 + G_{iij}^3\right) = 2g_i M_i^1 + D_{ii}^2 + C_{ii}^2,$$

where repeated indices imply summation. The unclosed spatial fluxes are approximated by a gradient-viscosity model found using the Chapman-Enskog expansion. These five equations can be written in terms of the disperse-phase density $\varrho_{\mathrm{p}} = \rho_{\mathrm{p}} M^0$, disperse-phase momentum $\varrho_{\mathrm{p}} \mathbf{U}_{\mathrm{p}} = \rho_{\mathrm{p}}(M_1^1, M_2^1, M_3^1)$, and granular energy $\varrho_{\mathrm{p}} e_{\mathrm{p}} = \frac{1}{2}\rho_{\mathrm{p}} M_{ii}^2 = \frac{1}{2}\rho_{\mathrm{p}}(M_{11}^2 + M_{22}^2 + M_{22}^2)$. In the very dilute limit, the disperse-phase viscosity and all other contributions due to collisions are negligible because they depend on $(M^0)^2 = \alpha_{\mathrm{p}}^2$ (see the definition of the linear collision model). Otherwise, for elastic collisions $C_{ii}^2 = 0$. Except for the drag terms (which are closed for Stokes drag (Marchisio and Fox, 2013)), the resulting system of equations is equivalent to the Euler equation of gas dynamics. In most gas-particle flow codes, the granular energy equation is replaced by an equivalent equation for the granular temperature. However, it is better to solve for the granular energy because it is a conserved quantity in the absence of drag and inelastic collisions. As is evident from Figure 8, the hydrodynamic model predicts so-called "sticky" particles (i.e., they accumulate at the mean velocity instead of crossing). In comparison, the Eulerian model with the QBMM closure allows the particle jets to cross, as is found using Lagrangian simulations. The crossing jets example clearly illustrates the importance of the closures used in moment methods, and the need to use closures that work over the entire range of possible flow regimes.

4.5 Boundary conditions and coupling with fluid solver

When using Eulerian moment methods, the issue of how to set the boundary conditions for the moments arises. With classical moment methods for which the reconstructed NDF is not available, it is difficult to fix the moments at boundaries. However, for QBMM for which the reconstructed NDF in the interior of the domain is known, any boundary conditions used in Lagrangian simulations can be used. In other words, if the fate of particles interacting with a boundary is known in the context of a Lagrangian simulation, exactly the same rule can be used with QBMM. For example, Maxwell proposed wall boundary conditions for the velocity NDF at a solid wall moving with zero velocity of the form

$$f_{\mathrm{w}}(\mathbf{v}) = \begin{cases} \xi f_{\mathrm{eq,w}}(\mathbf{v}) + (1 - \xi) f_{\mathrm{w}}^*(\mathbf{v}_{\mathrm{w}}) & \text{if } \mathbf{v} \cdot \mathbf{n}_{\mathrm{w}} > 0, \\ f_{\mathrm{w}}^*(-\mathbf{v}_{\mathrm{w}}) & \text{if } \mathbf{v} \cdot \mathbf{n}_{\mathrm{w}} \leq 0, \end{cases} \tag{105}$$

where $\mathbf{n}_{\mathrm{w}}$ is the outward-directed wall-normal vector, f_{w}^* is the (known) reconstructed velocity NDF at the wall, $\mathbf{v}_{\mathrm{w}} = \mathbf{v} - 2(\mathbf{v} \cdot \mathbf{n}_{\mathrm{w}})\mathbf{n}_{\mathrm{w}}$, and the accommodation coefficient is defined as follows.

- $\xi = 0$ yields specular reflections for which the velocity component normal to wall is reversed (i.e. $\mathbf{v} \to \mathbf{v}_{\mathrm{w}}$).
- $\xi = 1$ yields diffuse reflections for which particles are reflected with a given wall temperature and Maxwellian distribution $f_{\mathrm{eq,w}}$.

The top line in (105) corresponds to particles that move towards the wall and thus interact with it through the boundary condition, while the bottom line corresponds to particle moving away from the wall.

The velocity NDF f_{w} and f_{w}^* in (105) correspond to the values located exactly at the interface with the wall. In a finite-volume method, these values are not known but rather the cell-average NDF in the cell next to the wall n_1^n and in a ghost cell representing the wall n_0^n are known. It is thus necessary to compute the kinetics-based fluxes at the interface in a consistent manner by reconstructing the NDFs at the interface. For example, for a solid surface the mean wall-normal velocity at the wall should be null, which is equivalent to specifying that the mass flux across the interface is null. In terms of the cell-average NDFs, this constraint is

$$\int_{\mathbf{v} \cdot \mathbf{n}_{\mathrm{w}} > 0} |\mathbf{v} \cdot \mathbf{n}_{\mathrm{w}}| n_1^n \, \mathrm{d}\mathbf{v} = \int_{\mathbf{v} \cdot \mathbf{n}_{\mathrm{w}} < 0} |\mathbf{v} \cdot \mathbf{n}_{\mathrm{w}}| n_0^n \, \mathrm{d}\mathbf{v}, \tag{106}$$

which is exact if $n_0^n = f_{\mathrm{w}}$ from (105) with $f_{\mathrm{w}}^* = n_1^n$. More generally, any constraint used in the Lagrangian simulation must be consistent with the reconstruction of the cell-average wall NDF n_0^n at every time step n. Nonetheless, if the wall NDF is defined consistently with the Lagrangian model, the computation of the boundary conditions for QBMM and kinetics-based finite-volume methods is relatively straightforward.

The numerical algorithm for coupling the disperse-phase and continuous-phase solvers is described in detail in Passalacqua and Fox (2011a,b); Passalacqua et al. (2010). The transport step for the velocity moments is handled as described in Section 3. On the particle side, the drag force is modeled by

$$\mathbf{F}_{\mathrm{D}} = \frac{3m_{\mathrm{p}}\rho_{\mathrm{g}}}{4\rho_{\mathrm{p}}d_{\mathrm{p}}} C_{\mathrm{d}} \left|\mathbf{U}_{\mathrm{r}}\right| \mathbf{U}_{\mathrm{r}} \tag{107}$$

with drag coefficient

$$C_{\mathrm{d}} = \frac{24}{\mathrm{Re}_{\mathrm{p}}} \left(1 + 0.15\,\mathrm{Re}_{\mathrm{p}}^{0.687}\right). \tag{108}$$

The mass of a particle is m_{p} and the particle Reynolds number is defined by

$$\mathrm{Re}_{\mathrm{p}} = \frac{\rho_{\mathrm{g}}d_{\mathrm{p}}\left|\mathbf{U}_{\mathrm{r}}\right|}{\mu_{\mathrm{g}}}. \tag{109}$$

Note that $\mathbf{F}_{\mathrm{D},\alpha}$ is different for each abscissa because the relative velocity is defined by $\mathbf{U}_{\mathrm{r}} = \mathbf{U}_{\mathrm{g}} - \mathbf{v}_{\alpha}$. Since the drag force does not change the weights, i.e.,

$$\frac{\mathrm{d}n_{\alpha}}{\mathrm{d}t} = 0, \tag{110}$$

each velocity abscissa can be updated independently using

$$\frac{\mathrm{d}\mathbf{v}_{\alpha}}{\mathrm{d}t} = \frac{\mathbf{F}_{\mathrm{D},\alpha}}{m_{\mathrm{p}}}. \tag{111}$$

Note that other forces on the particles (e.g., lift, virtual mass, buoyancy, etc. (Marchisio and Fox, 2013; Vikas et al., 2011b)) can be treated in the same manner. At the end of the drag step, the optimal moment set is updated using the new values for the velocity abscissas.

On the gas side of the flow solver, the continuity equation (91) and momentum equation (92) with the momentum exchange term defined by

$$\boldsymbol{\beta}_{\mathrm{g}} = \frac{\boldsymbol{\beta}_{\mathrm{QMOM}}}{V_{\mathrm{p}}} \tag{112}$$

are solved using standard CFD algorithms. The total drag term for all abscissas is found from

$$\boldsymbol{\beta}_{\mathrm{QMOM}} = \sum_{\alpha=1}^{\mathcal{N}} n_{\alpha}\mathbf{F}_{\mathrm{D},\alpha}. \tag{113}$$

A partial elimination algorithm is used to improve the stability of the iterative solver when the drag coefficient becomes large. Overall, mass and momentum are fully conserved between the two phases. More details, along with applications of the gas-particle flow solver, can be found in Passalacqua and Fox (2011a,b, 2012); Passalacqua et al. (2010); Vikas et al. (2011a,b).

5　Conclusions

We conclude our presentation of quadrature-based moment methods and their application to polydisperse multiphase flows with a few comments on the extension to turbulent flows and some remarks concerning open questions.

5.1　Extension to turbulent flows

In this chapter, we have focused on using QBMM to develop mesoscale models for multiphase flows. Such models are equivalent to laminar flow models using in single-phase flows. When applied to turbulent multiphase flows, the computational cost of resolving all relevant scales is prohibitive, and alternative models are needed. In practice, Reynolds-average (RA) or large-eddy simulations (LES) are used to handle high-Reynolds-number flows.

LES/RA turbulence modeling of the mesoscale kinetic equation for gas-particle flow leads to

$$\frac{\partial \overline{n}}{\partial t} + \mathbf{v} \cdot \frac{\partial \overline{n}}{\partial \mathbf{x}} + \frac{\partial}{\partial \mathbf{v}} \cdot \overline{(\mathbf{A}\, n)} = \overline{\mathbb{C}[n, n]} \tag{114}$$

where the overbar indicates some type of averaging process that loses information about the unresolved scales. (In the following, we use RA quantities, but these could be replaced by filtered quantities for LES.) Observing this RA kinetic equation, we note the following.

- The free-transport term has same form as before and appears in closed form.
- QBMM can be directly applied to the RA moments of $\overline{n}$.
- Particle acceleration and collisions are unclosed and thus require turbulence models to close the terms involving fluctuations:

$$\overline{\mathbf{A}n} = \overline{\mathbf{A}}\,\overline{n} + \overline{\mathbf{A}'n'} \quad \text{RA fluid-velocity fluctuations}$$

$$\overline{\mathbb{C}} = \mathbb{C}[\overline{n}, \overline{n}] + \overline{\mathbb{C}'} \quad \text{RA particle-velocity fluctuations}$$

where $\overline{\mathbf{A}'n'}$ affects the single-particle acceleration model, while $\overline{\mathbb{C}'}$ is a second-order process that depends on two-particle statistics.

The current approach for modeling these terms is to modify existing Euler-Lagrange models by adding stochastic-based closures. (See other chapters in this book for examples.) However, due to the large range of flow conditions that can be modeled by the mesoscale kinetic equation (e.g., very dilute to dense granular flow), current turbulence closures are limited in their ability to extend beyond small ranges of flow conditions. It will therefore be necessary to revisit this research topic in order to develop more general closures.

An important technical point that has been completely overlooked in the literature on turbulence models for collisional gas-particle flows is the need to have separate models for the RA granular temperature and the turbulent kinetic energy of the disperse phase. In order to see why this is necessary, it suffices to consider the RA second-order velocity moments:

$$\overline{M}_{ij} = \overline{\int v_i v_j n \, d\mathbf{v}} = \int v_i v_j \overline{n} \, d\mathbf{v} \tag{115}$$

where the last equality follows from the linearity properties of Reynolds averages. Using the definition of the granular temperature, we can decompose M_{ii} as follows:

$$M_{ii} = M_{11} + M_{22} + M_{33} = \alpha_\mathrm{p}(|\mathbf{U}_\mathrm{p}|^2 + 3\Theta_\mathrm{p}). \tag{116}$$

The RA of this result is

$$\overline{M}_{ii} = \overline{\alpha_\mathrm{p}|\mathbf{U}_\mathrm{p}|^2} + 3\overline{\alpha_\mathrm{p}\Theta_\mathrm{p}} = \overline{\alpha}_\mathrm{p}\langle|\mathbf{U}_\mathrm{p}|^2\rangle_\mathrm{p} + 3\overline{\alpha}_\mathrm{p}\langle\Theta_\mathrm{p}\rangle_\mathrm{p} \tag{117}$$

where the second equality defines the phase averages $\langle|\mathbf{U}_\mathrm{p}|^2\rangle_\mathrm{p}$ and $\langle\Theta_\mathrm{p}\rangle_\mathrm{p}$. The phase-average mean square velocity can be further decomposed as

$$\langle|\mathbf{U}_\mathrm{p}|^2\rangle_\mathrm{p} = \langle|\langle\mathbf{U}_\mathrm{p}\rangle_\mathrm{p} + \mathbf{u}''_\mathrm{p}|^2\rangle_\mathrm{p} = \langle\mathbf{U}_\mathrm{p}\rangle_\mathrm{p} \cdot \langle\mathbf{U}_\mathrm{p}\rangle_\mathrm{p} + 2k_\mathrm{p} \tag{118}$$

where $\mathbf{u}''_\mathrm{p}$ is the fluctuating component of the disperse-phase velocity and $k_\mathrm{p} = \frac{1}{2}\langle|\mathbf{u}''_\mathrm{p}|^2\rangle_\mathrm{p}$ is the turbulent kinetic energy of the disperse phase. Combining these results yields

$$\overline{M}_{ii} = \overline{\alpha}_\mathrm{p}(\langle\mathbf{U}_\mathrm{p}\rangle_\mathrm{p} \cdot \langle\mathbf{U}_\mathrm{p}\rangle_\mathrm{p} + 2k_\mathrm{p} + 3\langle\Theta_\mathrm{p}\rangle_\mathrm{p}). \tag{119}$$

Because both k_p and $\langle\Theta_\mathrm{p}\rangle_\mathrm{p}$ appear in this expression, it is clear that knowledge of the trace of the RA velocity covariance is not sufficient to determine both of these quantities. In fact, it will be necessary to solve an additional turbulence model for k_p. In the literature on turbulent gas-particle flows, it is almost always assumed that k_p and $\langle\Theta_\mathrm{p}\rangle_\mathrm{p}$ are equivalent. Because the phase-average granular temperature appears in the linear collision model in the definition of the τ_C, this incorrect assumption will lead to an overprediction of the collision rate in turbulent flows.

5.2   Final remarks

Throughout this chapter we have attempted to provide the reader with a relatively comprehensive description of QBMM as applied to multiphase flow modeling. In summary, the principal points of interest are as follows.

- Mesoscale models have a direct link with the underlying physics and result in a kinetic equation for the one-particle NDF.
- QBMM solve the kinetic equation by reconstructing the distribution function from a finite set of its moments.
- NDF reconstruction algorithms based on quadrature require realizable moments.
- Numerical schemes (e.g. finite volume, time integration) used to solve the moment equations must ensure that the moments are always realizable.
- QBMM – combined with a fluid solver – handle fully coupled fluid-particle flows for all Stokes and Knudsen numbers.
- Because of the underlying physical model at the mesoscale, quadrature-based multiphase models are always mathematically well defined.
- Properly closed moment-based Eulerian models should agree with Lagrangian simulations that use the same mesoscale model.

Readers interested in more details on QBMM are referred to Marchisio and Fox (2013).

Because QBMM are relatively new, there are still many open questions. A short list of questions that need to be answered in the near future are as follows.

- Under what conditions will CQMOM yield unrealizable abscissas for bounded supports?
- Can we do better than extended CQMOM for multi-dimensional quadrature?
- Can we find better realizable high-order numerical fluxes for multivariate moments?
- When should we use QBMM instead of Lagrangian methods (or hybrid methods) to simulate a multiphase flow?
- Is it possible to find predictive turbulence models that are generally valid in the context of moment methods?

The reader can undoubtedly think of other important questions. However, given the generality and flexibility of QBMM, it is almost certain that answers to these questions will be found soon due to the growing number of researchers interested in this topic.

Acknowledgements

The author has been very fortunate to have had the opportunity to collaborate with a large group of excellent researchers, most of whom appear as coauthors in the references listed below. The author's work in polydisperse multiphase flow modeling using quadrature-based moment methods has been funded primarily by the U.S. Department of Energy – National Energy Technology Laboratory and the U.S. National Science Foundation. However, a small part of the research leading to the results on LES/RANS models has received funding from the European Union Seventh Framework Programme (FP7/2007-2013) under grant agreement No. 246556.

Bibliography

A. Buffo, M. Vanni, D. L. Marchisio, and R. O. Fox. Multivariate quadrature-based moment methods for turbulent polydisperse gas-liquid systems. *International Journal of Multiphase Flow*, 2012.

S. de Chaisemartin, L. Fréret, D. Kah, F. Laurent, R. O. Fox, J. Reveillon, and M. Massot. Turbulent combustion of polydisperse evaporating sprays with droplet crossing: Eulerian modeling and validation in the infinite Knudsen limit. In *Proceedings of the Summer Program 2008*, Center for Turbulence Research, Stanford, pages 265–276, 2008.

S. de Chaisemartin, L. Fréret, D. Kah, F. Laurent, R. O. Fox, J. Reveillon, and M. Massot. Eulerian models for turbulent spray combustion with polydispersity and droplet crossing. *Comptes Rendus Mécanique*, 337:438–448, 2009.

C. Chalons, R. O. Fox and M. Massot. A multi-Gaussian quadrature method of moments for gas-particle flows in a LES framework. In *Proceedings of the Summer Program 2010*, Center for Turbulence Research, Stanford, pages 347–358, 2010.

C. Chalons, R. O. Fox, F. Laurent, M. Massot, and A. Vié. A multi-Gaussian quadrature method of moments for simulating high-Stokes-number turbulent two-phase flows. In *Annual Research Briefs 2011*, Center for Turbulence Research, Stanford, pages 309–320, 2011.

J. C. Cheng and R. O. Fox. Kinetic modeling of nanoprecipitation using CFD coupled with a population balance. *Industrial & Engineering Chemistry Research*, 49:10651–10662, 2010.

J. C. Cheng, R. D. Vigil, and R. O. Fox. A competitive aggregation model for Flash NanoPrecipitation. *Journal of Colloid and Interface Science*, 351:330–342, 2010.

O. Desjardins, R. O. Fox, and P. Villedieu. A quadrature-based moment closure for the Williams spray equation. In *Proceedings of the Summer Program 2006*, Center for Turbulence Research, Stanford, pages 223–234, 2006.

O. Desjardins, R. O. Fox, and P. Villedieu. A quadrature-based moment method for dilute fluid-particle flows. *Journal of Computational Physics*, 227:2514–2539, 2008.

R. Fan, D. L. Marchisio, and R. O. Fox. Application of the direct quadrature method of moments to polydisperse gas-solid fluidized beds. *Powder Technology*, 139:7–20, 2004.

R. Fan and R. O. Fox. Segregation in polydisperse fluidized beds: Validation of a multi-fluid model. *Chemical Engineering Science*, 63:272–285, 2008.

R. O. Fox. *Computational Models for Turbulent Reacting Flows*. Cambridge University Press, Cambridge, UK, 2003.

R. O. Fox. Bivariate direct quadrature method of moments for coagulation and sintering of particle populations. *Journal of Aerosol Science*, 37:1562–1580, 2006.

R. O. Fox. CFD models for analysis and design of chemical reactors. *Advances in Chemical Engineering*, 31:231–305, 2007.

R. O. Fox. A quadrature-based third-order moment method for dilute gas-particle flows. *Journal of Computational Physics*, 227:6313–6350, 2008.

R. O. Fox. Higher-order quadrature-based moment methods for kinetic equations. *Journal of Computational Physics*, 228:7771–7791, 2009a.

R. O. Fox. Optimal moment sets for the multivariate direct quadrature method of moments. *Industrial & Engineering Chemistry Research*, 48:9686–9696, 2009b.

R. O. Fox. Large-eddy-simulation tools for multiphase flows. *Annual Review of Fluid Mechanics*, 44:47–76, 2012.

R. O. Fox, F. Laurent, and M. Massot. Numerical simulation of spray coalescence in an Eulerian framework: direct quadrature method of moments and multi-fluid method. *Journal of Computational Physics*, 227:3058–3088, 2008.

R. O. Fox and V. Raman. A multienvironment conditional probability density function model for turbulent reacting flows. *Physics of Fluids*, 16:4551–4565, 2004.

R. O. Fox and P. Vedula. Quadrature-based moment model for moderately dense polydisperse gas-particle flows. *Industrial & Engineering Chemistry Research*, 49:5174–5187, 2010.

L. Fréret, F. Laurent, S. de Chaisemartin, D. Kah, R. O. Fox, P. Vedula, J. Reveillon, and M. Massot. Turbulent combustion of polydisperse evaporating sprays with droplet crossing: Eulerian modeling of collisions at finite Knudsen and validation. In *Proceedings of the Summer Program 2008*, Center for Turbulence Research, Stanford, pages 277–288, 2008.

M. Icardi, P. Asinari, D. L. Marchisio, S. Izquierdo, and R. O. Fox. Quadrature-based moment closures for non-equilibrium flows: hard-sphere collisions and approach to equilibrium. *Journal of Computational Physics*, 231:7431–7449, 2012.

D. Kah, F. Laurent, L. Fréret, S. de Chaisemartin, R. O. Fox, J. Reveillon, and M. Massot. Eulerian quadrature-based moment models for dilute polydisperse evaporating sprays. *Flow, Turbulence, and Combustion* 85:649–676, 2010.

F. Laurent, A. Vié, C. Chalons, R. O. Fox, and M. Massot. A hierarchy of Eulerian models for trajectory crossing in particle-laden turbulent flows over a wide range of Stokes numbers. In *Annual Research Brief 2013*, Center for Turbulence Research, Stanford, pages 1–12, 2013.

N. Le Lostec, R. O. Fox, O. Simonin, and P. Villedieu. Numerical description of dilute particle-laden flows by a quadrature-based moment method. In *Proceedings of the Summer Program 2008*, Center for Turbulence Research, Stanford, pages 209–221, 2008.

Y. Liu, C. Cheng, R. K. Prud'homme, and R. O. Fox. Mixing in a multi-inlet vortex mixer (MIVM) for flash nano-precipitation. *Chemical Engineering Science*, 63:2829–2842, 2008.

D. L. Marchisio, A. A. Barresi, and R. O. & Fox. Simulation of turbulent precipitation in a semi-batch Taylor-Couette reactor. *AIChE Journal*, 47:664–676, 2004.

D. L. Marchisio and R. O. Fox. Solution of population balance equations using the direct quadrature method of moments. *Journal of Aerosol Science*, 36:43–73, 2005.

D. L. Marchisio and R. O. Fox. *Multiphase Reacting Flows: Modelling and Simulation*. Springer, Berlin, Germany, 2007.

D. L. Marchisio and R. O. Fox. *Computational Models for Polydisperse Particulate and Multiphase Systems*. Cambridge University Press, Cambridge, UK, 2013.

D. L. Marchisio, R. O. Fox, A. A. Barresi, M. Garbero, and G. Baldi. On the simulation of turbulent precipitation in a tubular reactor via computational fluid dynamics (CFD). *Chemical Engineering Research and Design*, 79:998–1004, 2001.

D. L. Marchisio, J. T. Pikturna, R. O. Fox, R. D. Vigil, and A. A. Barresi. Quadrature method of moments for population-balance equations. *AIChE Journal*, 49:1266–1276, 2004a.

D. L. Marchisio, J. T. Pikturna, L. Wang, R. D. Vigil, and R. O. Fox. Quadratic method of moments for population balances in CFD applications: Comparison with experimental data. *Chemistry and Engineering Transactions*, 1:305–310, 2004b.

D. L. Marchisio, R. D. Vigil, and R. O. Fox. Quadrature method of moments for aggregation-breakage processes. *Journal of Colloid and Interface Science*, 258:322–334, 2003.

M. Mehta, V. Raman, and R. O. Fox. On the role of gas-phase chemistry in the production of titania nanoparticles in turbulent flames. *Chemical Engineering Science*, 2012.

M. Mehta, Y. Sung, V. Raman, and R. O. Fox. Multiscale modeling of TiO_2 nanoparticle production in flame reactors: Effect of chemical mechanism. *Industrial & Engineering Chemistry Research*, 49:10663–10673, 2010.

A. Passalacqua and R. O. Fox. Advanced continuum modeling of gas-particle flows beyond the hydrodynamic limit. *Applied Mathematical Modelling*, 35:1616–1627, 2011a.

A. Passalacqua and R. O. Fox. Implementation of an iterative solution procedure for multi-fluid gas-particle flow models on unstructured grids. *Powder Technology*, 213:174–187, 2011b.

A. Passalacqua and R. O. Fox. Simulation of mono- and bidisperse gas-particle flow in a riser with a third-order quadrature-based moment method. *Industrial & Engineering Chemistry Research*, 2012.

A. Passalacqua, R. O. Fox, R. Garg, and S. Subramaniam. A fully coupled quadrature-based moment method for dilute to moderately dilute fluid-particle flows. *Chemical Engineering Science*, 65:2267–2283, 2010.

A. Passalacqua, J. E. Galvin, P. Vedula, C. M. Hrenya, and R. O. Fox. A quadrature-based kinetic model for dilute non-isothermal granular flows. *Communications in Computational Physics*, 10:216–252, 2011.

D. Piton, R. O. Fox, and B. Marcant. Simulation of fine particle formation by precipitation using computational fluid dynamics. *The Canadian Journal of Chemical Engineering*, 78:983–993, 2000.

V. Raman, H. Pitsch, and R. O. Fox. Eulerian transported probability density function sub-filter model for large-eddy simulations of turbulent combustion. *Combustion Theory and Modelling*, 10:439–458, 2006.

R. G. Rokkam, R. O. Fox, and M. E. Muhle. CFD modeling of electrostatic forces in gas-solid fluidized beds. *The Journal of Computational Multiphase Flows*, 2:189–205, 2010a.

R. G. Rokkam, R. O. Fox, and M. E. Muhle. Computational fluid dynamics and electrostatic modeling of polymerization fluidized-bed reactors. *Powder Technology*, 203:109–124, 2010b.

J. Sanyal, D. L. Marchisio, R. O. Fox, and K. Dhanasekharan. On the comparison between population balance models for CFD simulation of bubble columns. *Industrial & Engineering Chemistry Research*, 44:5063–5072, 2005.

S. T. Smith and R. O. Fox. A term-by-term direct numerical simulation validation study of the multi-environment conditional probability-density-function model for turbulent reacting flows. *Physics of Fluids*, 19:085102, 2007a.

S. T. Smith, R. O. Fox, and V. Raman. A quadrature closure for the reaction-source term in conditional-moment closure. *Proceedings of the Combustion Institute*, 31:1675–1682, 2007b.

M. Soos, L. Wang, R. O. Fox, J. Sefcik, and M. Morbidelli. Population balance modeling of aggregation and breakage in turbulent Taylor-Couette flow. *Journal of Colloid and Interface Science*, 307:433–446, 2007.

Y. Sung, V. Raman, and R. O. Fox. Large-eddy simulation based multiscale modeling of TiO_2 nanoparticle synthesis in turbulent flame reactors using detailed nucleation chemistry. *Chemical Engineering Science*, 66:4370–4381, 2011.

Q. Tang, W. Zhao, M. Bockelie, and R. O. Fox. Multi-environment probability density function method for modelling turbulent combustion using realistic chemical kinetics. *Combustion Theory and Modelling*, 11:889–907, 2007.

S. Tenneti, R. Garg, C. M. Hrenya, R. O. Fox, and S. Subramaniam. Direct numerical simulation of gas-solid suspensions at moderate Reynolds number: Quantifying the coupling between hydrodynamic forces and particle velocity fluctuations. *Powder Technology*, 203:57–69, 2010.

A. Vié, C. Chalons, R. O. Fox, F., Laurent, and M. Massot. A multi-Gaussian quadrature method of moments for simulating high Stokes number turbulent two-phase flows. In *Annual Research Brief 2012*, Center for Turbulence Research, Stanford, pages 309–320, 2012.

R. D. Vigil, I. Vermeersch, and R. O. Fox. Destructive aggregation: Aggregation with collision-induced breakage. *Journal of Colloid and Interface Science*, 302:149–158, 2006.

V. Vikas, C. D. Hauck, Z. J. Wang, and R. O. Fox. Radiation transport modeling using extended quadrature method of moments. *Journal of Computational Physics*, 2012a.

V. Vikas, Z. J. Wang, and R. O. Fox. Realizable high-order finite-volume schemes for quadrature-based moment methods applied to diffusion population balance equations. *Journal of Computational Physics*, 2012b.

V. Vikas, Z. J. Wang, A. Passalacqua, and R. O. Fox,. Realizable high-order finite-volume schemes for quadrature-based moment methods. *Journal of Computational Physics*, 230:5328–5352, 2011a.

V. Vikas, C. Yuan, Z. J. Wang, and R. O. Fox. Modeling of bubble-column flows with quadrature-based moment methods. *Chemical Engineering Science*, 66:3058–3070, 2011b.

C. Yuan and R. O. Fox. Conditional quadrature method of moments for kinetic equations. *Journal of Computational Physics*, 230:8216–8246, 2011.

C. Yuan, F. Laurent, R. O. and Fox. An extended quadrature method of moments for population balance equations. *Journal of Aerosol Science*, 51:1–23, 2012.

Y. Zou, M. E. Kavousanakis, I. G. Kevrekidis, and R. O. Fox. Coarse-grained computation for particle coagulation and sintering processes by linking quadrature method of moments with Monte Carlo. *Journal of Computational Physics*, 229:5299–5314, 2010.

A. Zucca, D. L. Marchisio, A. Barresi, and R. O. Fox. Implementation of the population balance equation in CFD codes for modelling soot formation in turbulent flames. *Chemical Engineering Science*, 61:87–95, 2006.

Mesoscopic particle models of fluid flows

Sauro Succi[*][‡]

[*] IAC-CNR, via dei Taurini, 19, 00185, Roma
Research Affiliate, Lyman Lab of Physics,
Harvard University, Cambridge (USA)

Abstract We review the general ideas behind coarse-grained representations of fluid dynamics, with special focus on two mesoscopic techniques which have proven particularly successful over the last two decades for the simulation of complex fluid flows, namely Dissipative Particle Dynamics and the Lattice Boltzmann method.

1 Introduction

Computational physics strives to imitate nature on a comparatively string-shoe budget: the gap between the degrees of freedom available to Nature and those affordable by our even most powerful foreseeable computers remains daunting, something of the order of the Avogadro number versus its square root, at best. Fortunately, Nature is kind to us and offers a huge amount of redundancy; very many of these degrees of freedom are not necessary, nor even desirable to know, to the purpose of understanding the actual behavior of fluids, and matter in general. This means that there is wide scope for devising stylized models of fluid behavior. Technically, the general procedure to remove redundant degrees of freedom goes by the name of *coarse-graining*, i.e. the process of distilling the essential physics, while relinquishing irrelevant details (1). A general procedure to perform such task in a systematic way is not available, and, strictly speaking, it might not even exist, since it does not correspond to any natural process: Nature does not need coarse-graining! As a result, coarse-graining is, to some extent, as much an art as a science.

The key issue is non-linearity and its distinctive property of transferring energy (and information) across different scales of motion. Because of this, upon coarse-graining, non-linear interactions generate new correlated interactions, which we shall call *scaling forces*, because they arise exquisitely from lack of scale invariance of non-linear interactions. As a general trend, coarse graining leads to modified forms of conservative interactions plus qualitatively new interactions, usually classified under the rubric of dissipative forces, typically representd by memory kernels and noise sources, the well-known Langevin approach. By design, coarse-graining leads to (many) less degrees of freedom, and under most circumstances, also more weakly interacting than the original ones. This is readily appreciated by considering that coarse-grained variables are "fatter", as they embody many microscopic degrees of freedom, and, by the very same reason, they interact on longer distances, hence more weakly. This is the very point of Renormalization-Group (RG) analysis; by recursive application of coarse-graining, one can hope that the interactions eventually become so weak that they can be handled perturbatively. However, this only works if correlations reorganize nicely, i.e. they lend themselves to a sensible regrouping within the same formal structure of the original ones, only with renormalized couplings. Unfortunately, this is not the rule, but a precious exception; usually correlations organize in a very tangled web of correlations, which does not fit within the same formal structure of the microscopic equations. That is why, the bottom-up (from micro to Macro) strategy to coarse-graining is often replaced and complemented by heuristic approaches, which postulate effective equations of motions based on general requirements of symmetry at the macroscopic level. Computational fluid dynamics falls plain within this framework.

Historically, the computational study of fluid-dynamic problems was almost invariably identified with the numerical solution of the Navier-Stokes equations of continuum mechanics. This is fairly reasonable, since continuum mechanics provides, in principle, the most economical description of the physics of fluids. A few continuum fields, density, pressure and the fluid velocity, provide a complete macroscopic description. Yet, several decades of continued efforts on the computational and analytical sides have taught us a hard-core lesson: despite their innocent-looking appearance, and very transparent physical meaning, the Navier-Stokes equations are exceedingly hard to solve. The reason, by and large, can be traced back to the strength of non-linear effects over dissipation, as measured by the Reynolds number, easily in the order of millions and above for most daily-life fluid phenomena, such as flows past cars and airplanes. The lesson taught by the Navier-Stokes equations is that the most economic level of representation from the conceptual viewpoint does not necessarily offer the most efficient computational scenario. Sometimes, coarse-grained equations turn out to be so complicated to solve, to the point of defeating the whole ordeal of having less degrees of freedom to solve for.

That is why coarse-graining is a fascinating and challenging task at a time.

2   Coarse-graining atomistic equations in space

Let us consider the equations of motion for a system of $i = 1, N$ classical point-particles (vector notation relaxed for simplicity):

$$\frac{dr_i}{dt} = v_i \tag{1}$$

$$m_i \frac{dv_i}{dt} = \sum_{j>i}^{N} f_{ij} \tag{2}$$

where $f_{ij} = f(r_i, r_j)$ are pairwise forces. Once the trajectory of each particle is known, based on initial and boundary conditions, this set of $6N$ discrete degrees of freedom can be organized in terms of continuum fields, such as the fluid mass density, momentum and energy (kinetic plus potential).

With reference to a region of space of volume $\Omega(x)$, centered about position x, we have:

$$\rho(x;t) = \tfrac{1}{\Omega(x)} \sum_{i=1}^{N} m_i S_i(x;t) \tag{3}$$

$$\rho(x;t)V(x;t) = \tfrac{1}{\Omega(x)} \sum_{i=1}^{N} m_i v_i S_i(x;t) \tag{4}$$

$$\rho(x;t)e(x;t) = \tfrac{1}{\Omega(x)} \sum_{i=1}^{N} \tfrac{m_i v_i^2}{2} S_i(x;t)) + \sum_{i=1;j>i}^{N} F_{ij}r_{ij}S_i(x;t)S_j(x;t) \tag{5}$$

where $S_i(x;t) \equiv S(x - r_i(t))$ is a suitable shape function, normalized to the total numbers of particles, i.e. $\sum_{i=1}^{N} S(x - r_i(t)) = N$. The simplest instance of shape-function is the Dirac-delta, but smoother options, such as piecewise linear or cubic functions, are better suited for numerical implementations. The Navier-Stokes equations need no more than the above: five continuum fields instead of Avogadro's numbers of particles! However, so far we have performed no coarse-graining at all; in fact, we have inflated the amount of information, because continuum fields contain in principle an infinite amount of degrees of freedom, since they are formally defined in the idealized limit $Vol(x) \to 0$. The actual reduction can only be appraised by attaching the fields to a discrete partition of space where no volume is allowed to shrink to zero size. A simple and natural way to coarse-grain this system is to partition the geometrical domain into a collection of subdomains Ω_I, $I = 1, N_B = N/B << N$.

The dynamical state of each subdomain (the mesoparticle, sometimes called simply *blob*) is characterized by its total mass, position and velocity, defined as follows:

$$M_I(t) = \sum_{i=1}^{N} m_i S_{iI}(t) \tag{6}$$

$$R_I(t) = \frac{1}{M_I} \sum_{i=1}^{N} r_i S_{iI}(t) \tag{7}$$

$$V_I(t) = \frac{1}{M_I} \sum_{i=1}^{N} m_i v_i(t) S_{iI}(t) \tag{8}$$

where $S_{iI}(t) \equiv S(r_i(t) - R_I)$. The next step is to devise the equations of motion for the mesoparticleis, based on the microscopic equations (1). By summing over all molecules in the mesoparticle, we obtain

$$\frac{dR_I}{dt} = (d/dt) \sum_i r_i S_{iI} = \sum_i (dr_i/dt) S_{iI} = \sum_i v_i S_{iI} = V_I \tag{9}$$

Note that $dS_{Ii}/dt = 0$ because the space derivative of the shape function is by design zero at $x = r_i$. Thus, the first equation of motion remains unchanged upon coarse-graining, which has to be the case, owing to its linearity. The momentum equation tells us a different story:

$$M_I \frac{dV_I}{dt} = (d/dt) \sum_i v_i S_{iI} = \sum_i \sum_j f_{ij} S_{iI}(t) S_{jJ}(t) \tag{10}$$

where we have taken $M_I = const = mB$ for simplicity. To compute the right hand side, it proves expedient to split the actual coordinates in terms of a mesoparticle coordinate plus a fluctuation around it, namely

$$r_{Ii} = R_I + \xi_{Ii}, \; i = 1, N_I \tag{11}$$

Note that the microscopic index i now runs locally in the I-th blob it belongs to. One can ithen expand the microscopic force around the barycentric coordinate $r = R_{IJ}$:

$$f_{ij} = f_{IJ} + f'_{IJ}\xi_{IJ,ij} + f''_{IJ}\xi^2_{IJ,ij} + f'''_{IJ}\xi^3_{IJ,ij} + \ldots \tag{12}$$

where primes stand for radial derivatives, divided by the corresponding factorial coefficient, and where we have set $f_{IJ} \equiv f(R_{IJ})$ and $\xi_{IJ,ij} = \xi_{Ii} - \xi_{Jj}$.

The force on the I-th blob writes as follows:

$$F_I = \sum_{i=1}^{N_I} \sum_{J \neq I}^{N_B} \sum_{j=1}^{N_J} \{f_{IJ} + f'_{IJ}\xi_{IJ,ij} + f''_{IJ}\xi^2_{IJ,ij} + f'''_{IJ}\xi^3_{IJ,ij} + \ldots\} \tag{13}$$

This also rewrites as:

$$F_I = \sum_{J \neq I}^{N_B} N_I N_J (f_{IJ} + g_{IJ}) \tag{14}$$

where we have set:

$$g_{IJ} = \frac{1}{N_I N_J} \sum_{i=1}^{N_I} \sum_{j=1}^{N_J} (f''_{IJ} \xi^2_{IJ,ij} + f'''_{IJ} \xi^3_{IJ,ij} + \dots) \equiv <f_{ij}> - f_{IJ} \tag{15}$$

The term g_{IJ} collects all the effects of the non-linear terms which do not cancel out upon coarse-graining. We shall call this scaling forces, since they arise exquisitely from the lack of scaling invariance of non-linear interactions.

Summarizing, the coarse-grained equations for the blobs take the final form

$$\frac{dR_I}{dt} = V_I \tag{16}$$

$$M_I \frac{dV_I}{dt} = \sum_{J \neq I} F_{IJ} = \sum_{J \neq I}^{N_B} N_I N_J (f_{IJ} + g_{IJ}) \tag{17}$$

Several remarks are in order.

First, we see that the blobs interact more weakly than the original molecules, because, after factoring out $M_I/N_I = m$, the coarse-grained forces F_{IJ} are weaker than the microscopic ones, f_{ij} simply because, on average, $R_{IJ} > r_{ij}$. Note indeed that intra-blob forces cancel out because of third Newton's law $f_{ij} + f_{ji} = 0$, so that the self-interaction term $J = I$, carrying the strongest interactions, is excluded by the summation.

This argument makes abstraction of the scaling forces. The latter, however, are expected to decay even faster with distance, because they are driven by higher order derivatives of the force. Thus, for decaying interactions of the form $f_{ij} \sim r_{ij}^{-\alpha}$, with $\alpha > D$ in D space dimensions (definition of short-range), the arguments holds solid. For long-range interactions, however, this is not necessarily true, because the cumulants $< \xi^p >_{IJ} \equiv \frac{1}{N_I N_J} \sum_i \sum_j \xi^p_{IJ,ij}$, $p = 2, 3...$ may diverge beyond a given p. This is the typical non-RG scenario, in which coarse-graining brings in non-ignorable terms at every iteration. Thus, we see that the coarse-grained equations, besides the original coarse-grain dof (R_I, V_I) also involve a bag of additonal fields, namely the cumulants of the relative positions within each blob. These are sometimes called auxiliary variables, and can be assimilated

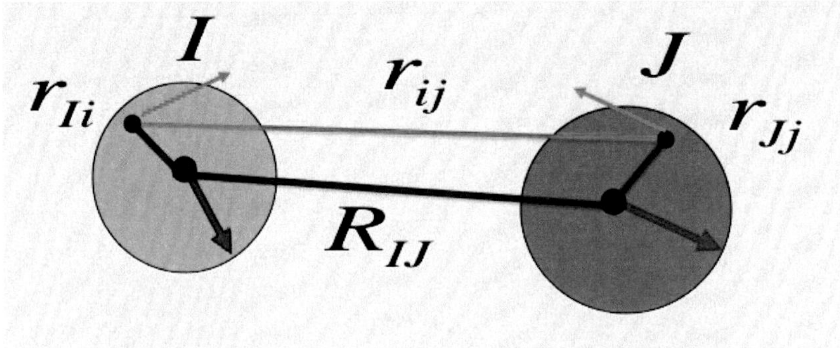

Figure 1. The basic splitting between fast and slow variables for spatial coarse-graining of atomistic dynamics

to a reservoir of fast degrees of freedom. The coupling to this reservoir, as we shall see, gives rise to qualitatively new effects, memory, noise and dissipation, which have no counterpart in the microscopic world. To better clarify this crucial item, it proves convenient to analyse the correlations arising by coarse-graining in time.

3   Coarse-graining in time

The correlation web emerging from coarse-graining in time can be illustrated for the case of a single degree of freedom, obeying Newton's equation:

$$m\ddot{x} = f(x) \tag{18}$$

Coarse-graining in time, over a time lapse T is defined as follows:

$$X(t) = \frac{1}{T} \int_{t-T/2}^{t+T/2} x(t')dt' \tag{19}$$

In the above, X is the slow variable (all modes with $\omega T >> 2\pi$ are filtered out) and $\xi = x - X$ is the fast degree of freedom, obeying $< \xi >= 0$ by definition.

The dynamic equations for the slow variable read as follows:

$$\dot{X} = V \tag{20}$$

$$m\dot{V} = f(X) + \sum_{n \geq 2} f^{(n)}(X) < \xi^n > \tag{21}$$

By subtracting the coarse grained equations from the microscopic ones, we obtain the equations of motion of the fluctuations, namely:

$$\dot{\xi} = \eta \tag{22}$$

$$m\dot{\eta} = (f - <f>) = f_1(X)\xi + \sum_{n \geq 2} f_n(X)(\xi^n - <\xi^n>) \tag{23}$$

where we have set $\eta \equiv v - V$. To be noted that the first derivative of the force at $x = X$, i.e. the second derivative of the potential, controls the linear stability of the fluctuations, as per the condition $f_1(X) < 0$. Depending on the strength of the remainder, this term may receive perturbative corrections, or, under less smooth circumstances, it might even overturn in sign, corresponding to the onset of local instabilities. As to the web of correlations, it is apparent from eq. (22) that the time derivative of the second order spatial cumulant, $<\xi^2>$, is driven by the mixed cumulant $<\xi\eta>$. The dynamic equation for the latter, on the other hand, involves cumulants of the form $<\xi f(X+\xi)>$, which give rise to a complex tangle of correlations at all orders, since $f(x)$ is generally a non-polynomial function of x. The dynamics of the hierarchy of correlations can be written in exact form as follows (2):

$$\dot{\mu}_{lm} = l\mu_{l-1,m+1} + m\sum_{k=1}^{\infty} f_k(X)\mu_{l+k,m-1} \tag{24}$$

where we have set $\mu_{lm} = <\xi^l\eta^m>$, the mixed cumulant of order (l,m). Note, that this is a system of first-order linear equations, whose coefficients depend non-linearly on the coarse-grain position X. From eq. (20), it is apparent that the slow motion is driven by the spatial cumulants $\mu_{l,0}$. The system (20), augmented with the equations of motion of the cumulants (24), is completely equivalent to the original problem (18). The solution to the cumulant dynamics in a time interval $[t, t + \Delta t]$, reads formally as follows:

$$\mu_{lm}(t + \Delta t) = e^{L_{lm,l'm'}(X(t))\Delta t} \mu_{l'm'}(t) \tag{25}$$

where $L_{lm,l'm'}(X)$ is the Liouville matrix of the cumulants, as defined by the rhs of the system (24). Note that, in the above, we have assumed that $X(t)$ does not vary appreciably on a time scale Δt, which is indeed the case by the very definition of X, so long as $\Delta t < T$. By inserting (25) in the time-discretized version of (20), we obtain a self-contained equation for the slow degree of freedom in each discrete interval $[t, t + \Delta t]$. Such a self-contained equation looks more cumbersome than the original one. The advantage, however, is that it is better suited to systematic approximations.

For instance, one could simply truncate the hierarchy by setting to zero the cumulants beyond a given order, so as to keep the Liouville matrix down to a low-order rank. A more sensible procedure is to appeal to an enslaving argument and express the higher-order cumulants via a closure relation of the form:

$$\mu_{lm}(t) = C_{lm}\mu_{20}(t)^{l/2}\,\mu_{02}(t)^{m/2} \tag{26}$$

and generalizations thereof (3). The validity of such a closure can be assessed as follows: i) Solve the original equation of motion (18), ii) Coarse-grain the outcome $x(t)$ to obtain $X(t)$ and iii) coarse-grain the cumulants according to their definition, namely, $\mu_{lm} = \frac{1}{T}\int_{t-T/2}^{t+T/2}(x(t')-X(t'))^l(v(t')-V(t'))^m dt'$.

Once this calibration is performed (say, by optimizing the coefficients C_{lm} in some suitable variational sense), the coarse-grained equations can then be solved using a time-step of the order of T, i.e. much larger than the one required to track the original system (taken 1 for convenience). To the best of the author's knowledge, such kind of "bootstrap" strategy has not received much attention in the current literature.

So much for coarse-graining in time.

When coarse-graining is performed in both space and time, the tangle becomes correspondingly more complicated, as each scaling force g_{IJ} depends on a *tower* of cumulants $\mu_{IJ,lm}$, thus a sequence of large arrays, of size $N_B \times N_B$, to be tracked in time. The information-compression factor, BT, can be as large as 10^{20}: from 10^{23} to 10^9 in space, and easily another 10^6 in time (from femtoseconds to nanoseconds). It is only natural that such an enormous compression cannot be bought for free! Details of the coarse-graining procedure matter a lot, and a variety of different techniques, using different shape functions, Voronoi tesselations and other technical devices, have been developed in the literature (5; 6), but none of them escapes the tangle, hence the need for some form of closure. Leaving aside details, which do matter, two qualitatively emergent effects of coarse-graining can be identified in full generality, namely *Non-locality* (in space and time), and *Dissipation*, which we now illustrate in a pedagogical form.

3.1 Memory and Dissipation: elimination of fast modes

Non locality is apparent through the chain of auxiliary equations; as is well known, a chain of first order ODE's is equivalent to a single ODE with memory kernel (7), a so called non-Markovian formulation. This is apparent

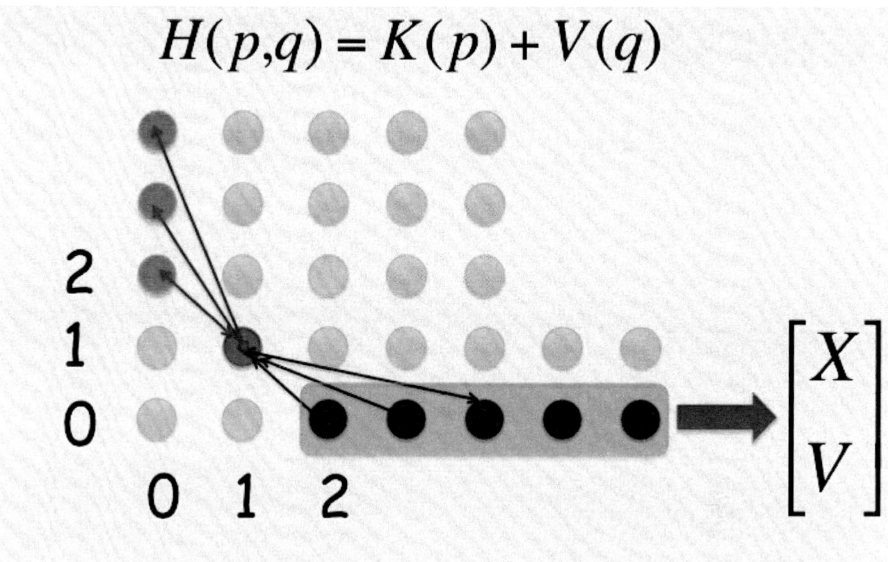

Figure 2. The tangled web of time-correlations for a general Hamiltonian of the form $H(p,q) = K(p) + V(q)$. Only the lowest-lying layer of cumulants couples to the coarse-grained variables (X, V). However, each cumulant couples to the others, according to the "butterfly" pattern depicted in the figure. In the figure we have indicated three levels of coupling along momentum (vertical axis), but for Newtonian mechanics they reduce to just one, because the kinetic energy is a quadratic function of the momentum. The coupling along position space can run to infinity whenever the potential $V(q)$ is a non-polynomial function of the position q.

already by a simple model 2×2 (linear!) problem

$$\frac{dx}{dt} = -ax + by \tag{27}$$

$$\frac{dy}{dt} = bx - cy \tag{28}$$

where we take $a, c > 0$ on account of stability. We shall further assume that $c >> a$, which configures y as the fast variable and x as the slow one. The basic idea is to eliminate the fast variable, to obtain an effective equation for the slow one. Solving the second, we obtain $y(t) = y(0)e^{-at} + b \int_0^t e^{-c(t-s)} x(s) ds$, which, upon insertion into the first, delivers an integro-differential equation for $x(t)$. The memory kernel $K(t, s) = e^{-c(t-s)}$ decays in a time lapse $1/c$, and can be replaced by a Dirac delta in the limit $|t-s| >> 1/c$, thereby returning the Markovian form $y(t) = bx(t)/c$, namely $\dot{x} = (-a + b^2/c)x$, i.e. a plain renormalization of the damping coefficient $a \to a - b^2/c$. To be noted that the effect of the eliminated variable is not exausted by the memory kernel, but includes the initial condition $y(0)$ as well. Thus, in order to obtain an effective equation for $x(t)$, an average over the initial conditions, $y(0)$, has to be taken too. Note that the effect of initial conditions fades away for $t >> 1/|c|$. This entails a loss of information, which is tantamount to irreversibility, hence dissipation. The toy model 27 is of course only a pale cartoon of the coarse-grained equations (16), but it illustrates nonetheless the two main points, namely i) a chain of ODE's is equivalent to a single non-Markov integro-differential iequation, with memory kernel, and ii) the elimination of fast variables, entails a loss of information which shows up as dissipation at a macroscopic level.

A physically more penetrating example is provided by the Brownian motion of a test particle in a bath of harmonic oscillators (7). The global system (test particle plus oscillators) is described by the overall Hamiltonian $H = H_1 + H_B$, where

$$H_1 = \frac{p^2}{2m} + V(x) \tag{29}$$

$$H_B = \sum_{j=1}^{N} \frac{p_j^2}{2} + \sum_{j=1}^{N} \frac{\omega_j^2}{2}(q_j - a_j x)^2 \tag{30}$$

In the above, the mass of the N harmonic oscillators has been set to 1 for simplicity, and the frequencies ω_j and amplitudes a_j are free-parameters. Since the coupling between the particle and the j-th oscillator in the bath is simply xq_j, the model can be solved exactly, to deliver the following analytical solution:

$$\frac{dp}{dt} = -V'(x) - \int_0^t K(t-s)p(s)ds + f_n(t) \tag{31}$$

where the memory kernel and the noise term are explictly given by ($\gamma_j = a_j\omega_j^2$):

$$K(t) = \sum_{j=1}^N a_j^2\omega_j^2 cos(\omega_j t) \tag{32}$$

$$f_n(t) = \sum_{j=1}^N \gamma_j p_j(0)t\frac{sin(\omega_j t)}{\omega_j t} + \sum_{j=1}^N \gamma_j(q_j(0) - a_j x(0))cos(\omega_j t) \tag{33}$$

It is thus appreciated that, by properly choosing the sequences a_j and ω_j, one can generate virtually any sort of memory kernel. It is also clear that the noise term is in fact perfectly deterministic, as long as the initial positions and momenta of the oscillators are known. Since N is a huge number, this is however unrealistic, and one is naturally led to average over the initial conditions. On the assumption that they obey Gaussian statistics, by the central limit theorem, their overall effect is assimilitated to an uncorrelated white noise. This is the assumption behind the emergence of the stochastic source in the Langevin equation.

This example is very precious, as it delivers an exact and very insightful expression of the memory kernel and stochastic source. However, it is perhaps even more for the limitations it exposes. namely that such insightful analytical treatment depends crucially on the linearity of the equations of motion. For the general case, there is no chance to extract such a clean insight; however, one can still appeal to universality arguments and postulate that some heuristic form of Langevin equation would capture the essence of the coarse-grained physics.

Coming back to the mesoparticle dynamics, the blob position-velocity (R_I, V_I) play the role of slow variables, while the auxiliary variables are the fast ones. The microscopic and coarse-grained representations, namely

$$\{r_i, v_i\} \rightarrow \{R_I, V_I, \mu_{IJ}^{lm}\}, \ I, J = 1, N_B, \ l, m = 1, 2 \ldots \tag{34}$$

are formally equivalent, as they contain the same amount of physical information. However, from the computational standpoint, they are surely not. The original microscopic system is under most circmusances, computationally unmanageable. The coarse-grained one, presents a tangle of space-time correlations which leaves little chance for exact analytical treatment. It does, however, open up plenty of room for informed approximations. Whence the motivation for heuristic models based on fictitious (meso)-particle dynamics.

4   Dissipative Particle Dynamics

Dissipative Particle Dynamics (DPD) was introduced in the seminal 1992 paper by Hoogerbrugge and Koelman (8), to cope with the main weaknesses of then existing mesoscopic methods, mainly Lattice Gas Cellular Automata and lattice Boltzmann, to be discussed shortly. The former was known to be faced with an exponential wall of computational complexity, while the latter was allegedly at odds with a proper account of thermal effects in fluids . The starting point of DPD is to formulate equations of motion for mesoscopic particles, each representing a large number of actual molecules, in terms of meso-particle centermass position and velocity, i.e a "blown-up" version of molecular dynamics. The distinctive feature of DPD is to acknowledge the presence of dissipative and random forces at the outset, and postulate the following heuristic equations of motion (vector notation reinstated):

$$\frac{d\vec{R}_I}{dt} = \vec{V}_I \tag{35}$$

$$M_I \frac{d\vec{V}_I}{dt} = \sum_J (\vec{F}_{IJ}^c + \vec{F}_{IJ}^d + \vec{F}_{IJ}^r) \tag{36}$$

where superscripts c, d, r, stand for conservative, dissipative and random, respectively.

The conservative forces read as follows:

$$\vec{F}_{IJ}^c = fW^c(R_{IJ})\hat{e}_{IJ} \tag{37}$$

where $\hat{e}_{IJ}$ is the unit vector along $\vec{R}_I - \vec{R}_J$ and $W^c(R_{IJ})$ is a shape function, giving the radial distribution of the conservative forces, with strength f. A popular choice is the piecewise linear function

$$W^c(r) = (1 - r/r_c)H(r - r_c) \tag{38}$$

where r_c is cutoff distance and H is the Heavyside distribution. More sophisticated inplementations make use of soft-core repulsive Lennard-Jones potentials. By soft-core, we imply that the interaction range scales up approximately as $r_c \sim (M/m)^{1/3}\sigma_m$, σ_m being the microscopic range of interaction between molecules of mass m, and M being the mass of the mesoparticle. With a blocking ratio $B = M/m \sim 10^{15}$, (Avogadro molecules versus, say, 10^8 dissipative particles), a naive spatial magnification gives $B^{1/3} \sim 10^5$, i.e. tens of microns, so that the system would reach up to centimeter size. The time magnification can be estimated as $B^{1/2} \sim 10^7$ leading from femtoseconds, the typical time-step in molecular dynamics, to tens of nanoseconds. Thus, a one-million DPD steps would span tens of milliseconds. So much for conservative interactions.

The friction forces reads as follows

$$\vec{F}_{IJ}^{d} = -\gamma W^d(R_{IJ})(\vec{e}_{IJ} \cdot (\vec{V}_I - \vec{V}_J))\hat{e}_{IJ} \qquad (39)$$

where γ is the friction coefficient, and the shape function $W^d(r)$ gives the interaction range of dissipative interactions. This is usually be taken in the form of a piecewise linear function, like $W^c(r)$, eventually raised to an integer power.

Finally, the random force is given by:

$$\vec{F}_{IJ}^{r} = \eta W^r(R_{IJ})\xi_{IJ}\hat{e}_{IJ} \qquad (40)$$

where $\xi_{IJ} = \xi_{JI}$ is a Brownian (Wiener) noise term, with non-zero average, and obeying the fluctuation-dissipation-theorem (FDT), $\xi_{IJ}(t)\xi_{KL}(t') = \eta_{IJ,KL}\frac{kT}{M}\delta(t - t')$.

In the above, η is the strength of the noise, $W^r(R_{IJ})$ gives the interaction range of random forces and $\eta_{IJ,KL}$ is the fourth-order viscous stress tensor. Compliance with the fluctuation-dissipation theorem implies $W^r(r)^2 = W^d(r)$. It is possible to show that the above set of stochastic ordinary differential equations (SODE's), conserve not only mass, but also total momentum. Consequently, they are capable of reproducing hydrodynamic motion at a macroscopic scale. In this sense, DPD can be regarded as a set of Langevin equations with momentum conservation.

4.1   Time-marching

Formally, the DPD model leads to a large set of SODE's, for which many time-integration schemes are available in the literature (9). A popular one is the dissipative-stochastic extension of the velocity Verlet scheme used in Molecular Dynamics (MD-VV). It reads as follows (superscripts $k = 1/2, 1$

denote the next half and full time-step, $t + k\Delta t$, respectively):

$$V^{1/2} = V + \frac{1}{2}\Delta V \tag{41}$$

$$R^1 = R + \frac{\Delta t}{2}V^{1/2} \tag{42}$$

$$V^1 = V + \Delta V(R^1, V^{1/2}) \tag{43}$$

In the above, ΔV is a short-hand for the velocity change in a time-step Δt, due to the acceleration F/M. It reads as follows:

$$\Delta V = \frac{1}{M}[(F^c + F^f)\Delta t + F^r(\Delta t)^{1/2}] \tag{44}$$

Note the Ito exponent $1/2$ which reflects the stochastic nature of the equation. Many variants have been discussed in the literature, which go beyond the scope of the present paper. The qualitative point to be made is that DPD time-marcher can borrow from the MD literature, but not import as-is from it, because of the friction and stochastic terms. The former requires particular care, since, at variance with standard Langevin formulations, the the friction force depends on both the relative velocity and position of the interacting particles, which implies some subtle modifications to the standard MD and Langevin integrators. It is worth to point out that, depending on the different type of time-integrator, the discreteness of the time-step Δt may have different implications for the energy equipartion and the structural properties of the DPD fluid.

4.2   Macroscopic limit

The DPD fluid can be shown to obey an equation of state of the form

$$P = \rho c_s^2 + P^{nid}(\rho) \tag{45}$$

where the sound speed is given by $c_s \sim \frac{\bar{r}}{\Delta t} < \xi >$, with $\bar{r} = \int W(r)rdr$ (Koelman and Hoogerbrugge take $W^d(r) = W^r(r)$), both piecewise linear with support r_c) and the non-ideal pressure is dictated by the conservative potential. The viscosity of the DPD fluid scales with

$$\nu \sim \gamma\frac{\bar{r}^2}{\Delta t} \tag{46}$$

with Δt a fraction of $t_c = r_c/c_s$. Apparently, these values are realized pretty closely (within percents) already with small ensembles of particles, i.e. for time scales not much longer than t_c and spatial scales not much larger than r_c, a nice feature which still begs for a clear theoretical explanation.

The statistical mechanics of DPD, as a genuine classical many-body system, was formally worked out by Espanol and Warren, who derived the associated Fokker-Planck equation and the explicit form of canonical equilibria (10). The same authors also pointed out a few interesting anomalies, notably lack of energy equipartition, due to the discreteness of the time step. In particular, they noted that even with a timestep of one thousands of r_c/c_s, the error $\delta T/T$ due to lack of equipartition is still in the order of a few percent, indicating the need for recalicabrating the temperature, $T \to T + \delta T$. A rigorous bottom-up approach, starting from the atomistic equations of motion, was subsequently developed by Flekkoy and Coveney, by means of an advanced coarse-graining procedure based on (dynamic) Voronoi tesselation (4). This has several appealing properties, including the potential capability of adjusting the mesoparticle size dynamically in space-time, depending on the actual resolution demand: a natural form of adaptive computing. On the critical side; from the theoretical viewpoint, the rigorous bottom-up procedure still needs to invoke some heuristic closure on the stress tensor (viscosity). From the practical one, the use of adaptive Voronoi tesselations in three-dimensions faces with a major computational burden. This is still very much an open front of basic and applied research in the field.

5 DMD at work

DMD can achieve major computational savings over molecular dynamics. These can be estimated as follows. Space is upscaled by a factor $B^{1/3}$, while, with diffusive processes in mind, time upscales by a factor $B^{2/3}$. As a result the space-time volume of the simulation, $V_4 = L^3 T$, grows by a factor $B^{5/3}$. With an algorithmic complexity scaling like V_4^a, and an exponent a between 1 and 2, we obtain a DPD/MD gain of about $G = B^{5a/3}$. With $a = 1$ (fully local algorithm) and $B = 10^3$, this gives $G = 10^5$. If, on the other hand, $a = 2$ (fully non-local), then, even a mild blocking ratio $B = 10$ already gives $G \sim 2 \, 10^4$. Thus, computational savings are potentially conspicuous.

However, in order to realize them in actual practice, one needs to make sure that the DPD model is capable of reproducing the major equilibrium and non equilibrium properties of coarse-grained fluids, namely, the equation of state, compressibility factors, diffusivity, viscosity and other transport coefficients. To this purpose, a validation-calibration procedure is required, before DMD can be operated at larger scales. This procedure consists of the following three basic steps:

1. Run nanoscale MD simulations

2. Coarse grain the MD data to produce CGMD (Coarse-Grained MD)

datasets for different values of the blocking ratio B.

3. Run DMD and fine-tune the DMD parameters so as to minimize the discrepancy between DPD and CGMD data, across the explored range of B values.

Once the above procedure is completed, one can finally apply DMD to larger scale problems (the validation procedure is usually performed with no more than a few thousands particles, on volumes of the order of $(10r_c)^3$). The DMD free parameters are the strengths f and η of the conservative and random interactions, respectively, and eventually the shape functions $W^c(r)$ and $W^f(r)$ as well. Further tuning parameters are often employed in the time-marching scheme (a not-so elegant, yet pretry effective, feature).

A full match between CGMD and DPD across the whole set of target observables is hardly achieved, but partial matches are acceptable, depending on the specific problem at hand. A particularly remarkable case of successful match is reported for the case of the Flory-Huggins model of polymer flows (11).

5.1 DPD summary

Summarizing, DPD is explicitly patterned after coarse-graining of molecular dynamics, although, failing a rigorous coarse-graining, it only retains the essential features: friction, noise and dissipation (in Markov form). In principle, one could also account for memory effects, by turning the friction coefficient into a memory kernel (non-Markovian DPD). However, this entails a significant computational burden, and to date, standard DPD typically comes in non-Markovian form. Thus, despite its explicit particle-like format, DPD is most appropriately regarded as a top-down scheme, i.e. a microscopic formalism designed based on macroscopic requirements. Nevertheless, compliance with fluctuation-dissipation theorem, as combined with momentum conservation, confers DPD the capability to account for thermal fluctuations, which is of great value in fluid simulations at the micro and nanoscale. Another practical plus is the fact of looking formally very similar to molecular dynamics, which permits to import a wide array of numerical techniques developed by the MD community for about half a century. On the minus side, the difficulty of realizing small viscosities, a common problem of particle methods based on direct interactions, either via soft-core potentials or hard-core collisions. To date, DPD has found significant use for the simulation of soft-matter systems, such as amphiphilic mixtures, polymer melts and similar. As a side-bonus, DPD is often used as a momentum-conserving thermostat for molecular dynamics simulations. For a review see (12).

6 Lattice Gas Cellular Automata

As noted above, DPD is explicitly patterned after molecular dynamics, in the sense that it retains in full the notion of particles trajectories evolving under the action of smooth potentials, plus non-smooth stochastic sources. However, other forms of coarse-graining are available, which appeal to the notion of *probability distribution functions*, rather than particle trajectories. In particular, one can sit in a given cell of phase-space, and keep record of the statistics of time-evolving populations in that cell: an Eulerian versus Lagrangian approach in phase-space.

Perhaps, the most radical instance of this statistical coarse-graining is provided by Lattice Gas Cellular Automata (LGCA) (13).

The crucial idea is to formulate a fictitious particle dynamics, disregarding irrelevant microscopic details except those that are essential to preserve the basic symmetries of the equation of motion of fluids. A strong call to universality.

LGCA came to a pinnacle of this idea by formulating the pseudo-particle dynamics in *Boolean algebra*! Dissipation is not built-in, but emergent from elimination of high-order kinetic moments via the assumption of weak departure from local equilibrium. The basic objects are the occupation numbers $n_i(x; t)$, which take value 1 or 0 depending on whether a particle sitting at lattice site $\vec{x}$ at time t with discrete speed $\vec{c}_i$, is there or not. The discrete speeds are identified with the links connecting each lattice sites to its b neighbors $\vec{x}_i = \vec{x} + \vec{c}_i$. Being a pure yes/no quantity, n_i, can be encoded within a single bit: a boolean variable. Quite remarkably, provided the lattice affords sufficient symmetry, the basic equations of fluids, as expressed by mass, momentum and energy conservations, as well as rotational invariance, can be encoded within a fully boolean equation of motion of the form:

$$n_i(\vec{x} + \vec{c}_i, t + 1) - n_i(\vec{x}, t) = C_i[n] \qquad (47)$$

In the above, the left hand side stands for the free motion of the particles, whereas the right hand side describes their interactions, in the form of local multi-body collisions. The specific form of the collision operator C_i can be rather cumbersome, but its essential properties are not,

$$\sum_i C_i\{1, \vec{c}_i, c_i^2\} = \{0, \vec{0}, 0\} \qquad (48)$$

These can be secured compatibly with the constraint $C_i = \{-1, 0, 1\}$, which preserves the boolean nature of the occupation numbers.

A typical boolean collision in the prototypical hexagonal lattice, may look like follows:

$$C(1, 4 \rightarrow 2, 5) = n_1 n_4 (1 - n_2)(1 - n_5), \tag{49}$$

This expression is readily checked to take the value 1 if and only if there are two head-on particles moving right and left (1 and 4) prior to the collision and one right-east (2) and south-west (5), afterwards. For any other configuration, $C = 0$. The collisional contribution $+1$ is then added to states 2 and 4, subtracted to states 1 and 2, while leaving all others unchanged. It is readily checked that this is obtained by applying the "exclusive OR" (XOR) operator to the string of bits representing the state of the automaton. Thus, the post collisional state is simply $n' = (XOR)\, n$. Each potential collision turning the pre-collisional configuration s (a string of 6 bits in the hexagonal lattice), into the post-collisional s', comes with its own operator $C(s \rightarrow s')$ attached.

How does the LGCA reproduce the true equations of fluids?

The procedure is basically the standard Chapman-Enskog asymptotics, as adapted to the lattice. This is widely documented in the literature, so that here we shall just outline the basic conceptual steps. First one defines the probability distribution function

$$f_i(\vec{x}; t) = < n_i(\vec{x}; t) > \tag{50}$$

namely, the probability (density) of finding a particle at site $\vec{x}$ ad time t, with discrete speed $\vec{c}_i$. In the above, brackets stand for ensemble averaging over microscopic configurations. One readily recognizes f_i as a lattice analogue of the Boltzmann distribution $f(\vec{x}, \vec{v}; t)$. With f_i at hand, all fluid quantitites of interest, such as the mass, momentum and energy, follow by linear combinations of the discrete velocity states (latin subscripts label spatial dimensions):

$$\rho(\vec{x}; t) = m \sum_i f_i(\vec{x}; t) \tag{51}$$

$$\rho u_a(\vec{x}; t) = m \sum_i f_i(\vec{x}; t) c_{ia} \tag{52}$$

$$\rho e(\vec{x}; t) = \frac{m}{2} \sum_i f_i(\vec{x}; t) c_i^2 \tag{53}$$

$$\tag{54}$$

The discrete Boltzmann distribution is postulated to obey the analogue of the LGCA microdynamics, namely:

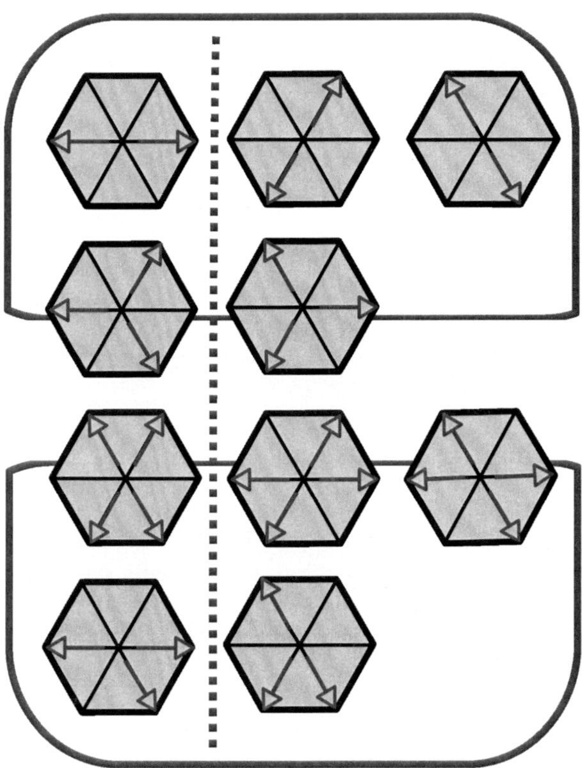

Figure 3. A set of possible collisions in the FHP hexagonal lattice. Left: pre-collision, right: post-collision.

$$f_i(\vec{x} + \vec{c}_i, t + 1) - f_i(\vec{x}, t) = C_i[f] \tag{55}$$

This is of course true only upon neglecting many-body correlations, the usual molecular-chaos assumption. Next one makes the assumption that f never departs too strongly from the local equilibrium, f_i^{eq}, the discrete analogue of the local Maxwell-Boltzmann equilibria, dictated by the condition $C[f^{eq}] = 0$. By writing $f = f^0 + \epsilon f^1$, with ϵ a smallness parameter of the order of the Knudsen number, i.e. the ratio between the molecular mean free path and a typical scale of change of fluid quantities, the Chapman-Enskog procedure permits to derive the Euler equations of inviscid fluid to order zero, and the Navier-Stokes equations of dissipative fluids at order one. This makes a long story very short, but since all details are fully documented in the vast literature on the subject, here it is sufficient to provide only a few general considerations. The LGCA promises revolutionary scenarios, mostly on account of its Boolean nature and the resulting freedom from round-off errors, plaguing floating-point simulations. However, a closer inspection revealed a number of practical issues, such as i) statistical noise (Boolean systems need massive averaging to be produce smooth signals) and ii) unsustainable (exponential) complexity of the collision operator in three dimensions. In fact, the operator C requires of the order of 2^b boolean operations at each lattice site and time instant, and since $b = 24$ in $D = 3$, this makes about 16 Million operations per lattice sites and time-step. And this for just one fluid! Among others, this major weakness thwarted the development of LGCA and its application to high-Reynolds flows.

7 Lattice Boltzmann

Historically, LB was born on the wake of the noise problem of LGCA's. However, in the subsequent years, it rapidly evolved into a self-standing method, with applications throughout fluid dynamics at virtually all scales of motion (16). Here again, we only give the essentials. The first LBE takes exactly the form (55), i.e. a pre-averaged LGCA. By definition, this form faces the very same computational wall experienced by LGCA, hence it is equally unviable for practical computations in three dimensions. In the subsequent years, a small of group of researchers realized that by appealing again to the notion of weak departure from local equilibria, the collision operator could be brought down to a simple matrix-relaxation around local lattice equilibria.

The resulting LBE reads as follows:

$$f_i(\vec{x} + \vec{c}_i, t + 1) - f_i(\vec{x}, t) = \Omega_{ij}(f_j^{eq} - f_j)(\vec{x}; t) \tag{56}$$

This makes a world of a difference, since the rhs can now be computed with order b^2 operations, instead of 2^b!. The relaxation matrix Ω_{ij} was initially derived directly from LGCA rules, i.e the boolean collision matrix $C(s, s')$. This results in a severe penalty on the number of collisions allowed by the scheme, hence on the minimum viscosity which can be attained on the lattice (in which many less collisions are allowed, due to conservation constraints). We remind that viscosity, namely momentum diffusivity, scales with the molecular mean-free path, which, in turn, decreases with increasing collisionality. Hence small viscosities can only be attained by performing many collisions per time-step.

This problem was lifted by realizing that the relaxation matrix can be designed top-down, solely based on symmetry requirements of the Navier-

Stokes equations. By doing so, the viscosity can be brought as low as allowed by the lattice resolution and not by inherent collisional constraints. In other words, hydrodynamics settles down at the level of a single lattice cell, which was not the case for LGCA, nor for any previous LBE. This marks a very distinctive point for LB versus other particle methods: the mean free path is no longer dictated by the actual collisions that can be realized in the lattice, but can be chosen at the outset as a free parameter, i.e. by prescribing the rate of convergence to local equilibrium. Of course, since the equilibrium is no longer attained through explicit collisions, this top-down approach is exposed to a realizability issue, which we shall comment upoin shortly.

The final step in the development of basic LBE theory was brought about by the realization that the relaxation matrix can be taken in diagonal form, giving rise to the so-called lattice BGK (Bhatnagar-Gross-Krook) scheme, by far the most popular form of LBE to this date.

It reads as follows (17):

$$f_i(\vec{x} + \vec{c}_i, t + 1) - f_i(\vec{x}, t) = -\omega(f_i - f_i^{eq})(\vec{x}; t) \tag{57}$$

In this version, all symmetries securing compliance with macroscopic hydrodynamics are in charge of local equilibria, which must obey the following mass, momentum and momentum-flux conservations:

$$\sum_i f_i^{eq} = \rho \tag{58}$$

$$\sum_i f_i^{eq} c_{ia} = \rho u_a \tag{59}$$

$$\sum_i f_i^{eq} c_{ia} c_{ib} = \rho(u_a u_b + c_s^2 \delta_{ab}) \tag{60}$$

where latin indices run over spatial dimensions.

A suitable expression fulfilling these rules is:

$$f_i^{eq} = \rho w_i (1 + u_i + q_i) \tag{61}$$

where $u_i = u_a c_{ia}/c_s^2$ and $q_i = (u_i^2 - u^2/c_s^2)/2$. In the above, $c_s^2 = \sum_i w_i c_{ia}^2$, is the lattice sound speed, and w_i is a set of lattice-specific weights normalized to unity.

Note that these equilibria are second-order expansions in the Mach-number $M = u/c_s$, of the continuum Maxwell-Boltzmann distribution. Indeed, the mere replacement $v_a = c_{ia}$ in Maxwell-Boltzmann distribution would -not- and can-not fulfill the constraints (58), for any generic value of the fluid speed. The reason is basic: Maxwell-Boltzmann equilibria are gaussian, hence correspond an infinite series in powers of the Mach number u/c_s, the coefficient of the n-th term, being a rank n Hermite tensor. By construction, the standard LB lattices ensure isotropy to 4th order only, which means that higher order terms would break isotropy and Galilean invariance. Going to higher speeds and/or thermal fluids, is of course possible, but necessarily involves the use of higher order lattices (See section on thermal flows).

The eq. (57) describes a fluid with equation of state $P = \rho c_s^2$ and a viscosity

$$\nu = c_s^2 \left(\frac{1}{\omega} - \frac{1}{2} \right) \tag{62}$$

in lattice units, $\Delta x = \Delta t = 1$.

A very remarkable property of this expression, which proves crucial for high-Reynolds applications, is the $-1/2$ shift, which permits to realize small viscosities, $\nu = O(\epsilon)$, with time-steps still order $O(1)$, by choosing $\omega \sim 2 - \epsilon$. Note that, unlike the original LB in pre-averaged format, here collisions are not performed directly, but are expressed instead as a relaxation around a -given- local equilibrium. This cuts down dramatically the computational cost, and permits to reach very low viscosities, but also exposes LB to numerical instabilities, due to the potential lack of an underlying H-theorem (microscopic realizability). It is a very lucky instance that the top-down LB can work at very-low viscosities, typically of the order of $1/N$ with N lattice sites per dimension, before such instabilities become manifest. Most importantly, in the subsequent years, LB versions compliant with the H-theorem and the entropic principle, have been discovered and vigorously developed (18). This marked a major progress in the LB theory, as it closed the realizability gap opened up by the top-down approach.

The ability to reach down very low viscosities confers LB its unique capability of straddling across virtually all scales of motion, from nanofluids all the way up to very high-Reynolds turbulent flows completely out of reach for particle methods.

7.1   Non-ideal fluids

Another major asset of LB is its flexibilty towards the inclusion of additional interactions, on top of plain fluid-dynamics. We are referring to the interaction term $S(x,v) \equiv (F(x)/m)\partial_v f$ in the continuum Boltzmann equation, expressing the coupling of the fluid to an external (or internal) force field $F(x)$. This can either represent an external field, such as gravity, or more complex interactions, such as the effective forces resulting from

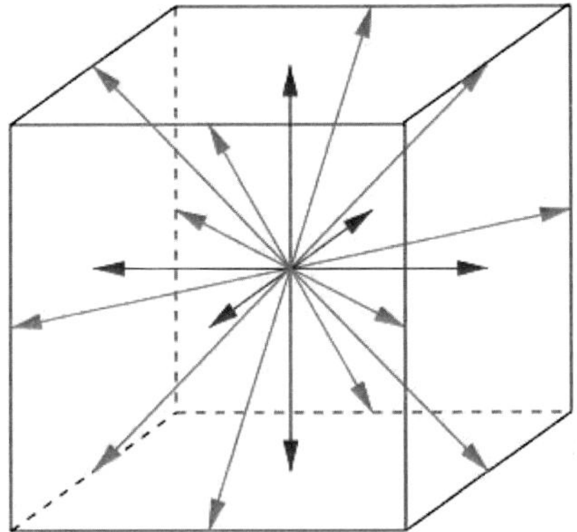

Figure 4. The basic set of discrete velocities for three-dimensional LB simulations. These include one zero-speed rest particle (center), six speed-1 face-center, and twelwe speed-2 edge-centers, for a total of nineteen (D3Q19)

intermolecular interactions, eventually driving dynamic phase-transitions (gas-liquid) within the flow.

To this regard, a particularly successful LB model for non-ideal fluids, is the one due to Shan-Chen (SC) (19). In this model, the interaction force is taken in the form

$$F_a(x) = \psi[\rho(x)] \sum_i w_i G_i c_{ia} \psi[\rho(x + c_i)] \tag{63}$$

where subscript a runs over spatial dimensions. In the above, $\psi[\rho]$ is a generalized density, reducing to the physical one in the limit $\rho \to 0$, and G is the strength of the non-ideal interactions. In the original SC model the sum extends over the first Brillouin region, with $G_i = G$, although multi-range generalizations have recently proven to encode a very rich phenomenology, relevant to soft-glassy materials, such as foams and emulsions. It is clear that (63) is a lattice analogue of the effective one-body force, with $G_i \equiv G(c_i^2)$ playing the role of the two-body radial correlation. The original SC model only caters for attractive interactions, $G < 0$, and can be shown to contribute a non-ideal pressure:

$$P^*(\rho) = \frac{G}{2} c_s^2 \psi^2(\rho) \tag{64}$$

With the choice $\psi = 1 - e^{-\rho}$, this can be shown to support a phase-transition at $\rho_{crit} = log2$ for $G < G_{crit} = -4$. The beauty of this model is that the non-ideal interaction is readily introduced with a few lines of code, by computing the bias on the i-the population due to the interaction term, namely:

$$S_i = w_i(S_a^1 c_{ia} + S_{ab}^2(c_{ia}c_{ib} - c_s^2\delta_{ab}))$$ (65)

where S_a^1, S_{ab}^2 are the mass, momentum and momentum-flux input rates per unit time due to the force term.

This source simply adds to the rhs of the LBE.

The SC model shines for its conceptual transparence and computational efficiency. Of course, this does not make it immune from criticism, primarily the existence of spurious currents near interfaces with strong density-contrasts, which hamper stability for density jumps beyond 1 : 50. Many remedies have been proposed in the recent literature, and make the object of an intense activity in modern LB research.

The SC model and extensions thereof are currently used for a wide range of complex multiphase and multicomponent fluid simulations.

7.2   Thermal flows

The simulation of thermal phenomena with lattice fluids meets with some conceptual and practical subtleties.

Indeed, the central idea of LGCA and LBE methods is to constrain the molecular velocities within a given crystallographic structure, the lattice. Since these velocities are "frozen", with no dispersion around them, the very notion of temperature, as we know it from continuum kinetic theory, has to be handled with great care. In fact, strictly speaking, it simply does not apply. Indeed, the original LB method was intended mostly for iso-thermal (or better yet, a-thermal) flows, in which the temperature is just a control parameter, with no dynamic meaning. In fact, for ideal fluids, the thermal speed identifies with the sound speed, $\frac{kT}{m} \equiv v_T^2 = c_s^2$, the latter being fixed by the discrete lattice speeds. In order for the temperature to acquire a dynamical status, further macroscopic quantities need to be recovered within the lattice structure. More precisely, third order kinetic moments (heat flux), which require isotropic sixth-order lattice tensors of the form $T_{abcdef} = c_{ia}c_{ib}c_{ic}c_{id}c_{ie}c_{if}$. This calls for higher order lattices, extending beyond the first Brillouin cell, i.e. with more than $3^3 = 27$ neighbors. The corresponding lattice thermal equilibria read as follows

$$f_i^{eq} = \rho w_i(1 + u_i + q_i + t_i)$$ (66)

where $t_i = c_{ia}c_{ib}c_{ic}u_a u_b u_c - c_s^2 P[u_a \delta_{bc}]$ and P means permutations.

The thermal speed, defined by:

$$\rho v_T^2 = \sum_i f_i (c_i - u)^2 \tag{67}$$

and the corrsponding heat flux

$$q_a = \sum_i f_i (c_{ia} - u_a)(c_i - u)^2 \tag{68}$$

acquire a dynamic meaning, and the lattice fluid is freed from the isothermal constraint $v_T^2 = c_s^2$. This, however does not full circle the story. In fact, stability issues remain open, as signalled by the strong numerical instabilities which develop as soon as one tries to insert the dynamic thermal speed, v_T, in the local equilibria, instead of the static one, c_s. This is no accident, since the dynamic temperature, defined through (67), coincides with the equilibrium temperature (the one appearing in the Maxwell-Boltzmann equilibria), only if local equilibria are Maxwellian, which is not the case in the lattice. Several recipes have been devised to cope with this problem. The earliest thermal LB models just made use of higher order lattices, to match third order kinetic moments (20). As hinted above, they face severe stability problems as soon as the temperature departs significantly from the reference value $kT_0/m = c_s^2$. Subsequently, it was shown that the stability problem can considerably be alleviated by introducing two separate discrete distributions, one for the ordinary mass and momentum carrier, and one for the energy and energy-flux carriers (21). The advantage of using two distinct populations is that none of them has to match sixth order tensors, so that ordinary isothermal lattices can be employed and stability issues considerably mitigated. The price is doubling of the degrees of freedom. To be noted, that even in this case, substitution of the dynamic temperature in the local equilibria leads to stability problems. Finally, a third, very pragmatic solution, is to solve the temperature equation by an independent method, typically finite-differences, while leaving the LB treatment only for the fluid. Each of these options has merits and downsides, but it appears like the double-distribution approach is presently the most popular one. To the best of the author's knowledge, despite significant progress (22), a fully consistent thermal LB scheme, meaning by this one in which the local equilibrium responds to the dynamic temperature, remains to be developed.

7.3   Fluctuating LB

Under the relentless drive of moder science and technology and miniaturization, there has a growing motivation to explore fluid behavior at the micro and nanoscale. Ironically, this has prompted the need for reinstating the effect of fluctuations, whose elimination was at the origin of LB itself!. The need for accounting for fluctuations is readily appreciated by noting that each LB particle corresponds to

$$B_{lb} = \frac{n\Delta x^3}{n_{lb}} \tag{69}$$

fluid molecules. In the above, n and n_{lb} are the fluid density in physical and LB units respectively. The customary choice is $n_{lb} = 1$. This is the blocking factor of LB simulations.

For macroscopic flows, say $\Delta x = 1$ mm, this is indeed a huge number, easily in order of 10^{20}. However, things change as the nanoscale is approached. With a reference mesh spacing $\Delta x = 1$ nm, a typical fluid, say water, would yield $B_{lb} = 30$, which gives a signal/noise ratio of the order of $1/\sqrt{30} \sim 0.2$. This is somewhat pessimistic, since one should also consider time-averaging, but by and large, it shows that taking LB applications below the nanometer scale should handled with great caution. Trailbalzing work in this direction was performed by A. Ladd (23), who introduced a fluctuating Lattice Boltzmann (FLB) by making an explicit connection with Landau-Lifshitz fluctuating hydrodynamics (24). The basic idea is to introduce a fluctuating source term in the form:

$$\tilde{S}_i = w_i \tilde{S}_{ab} Q_{iab} \tag{70}$$

where the fluctuating stress tensor obeys the tensor FDT relation:

$$\tilde{S}_{ab}(\vec{x};t)\tilde{S}_{cd}(\vec{x}',t') = \frac{kT}{m}\eta_{ab,cd}\delta(\vec{x} - \vec{x}')\delta(t - t') \tag{71}$$

where $\eta_{ab,cd}$ is a fourth-order isotropic tensor parametrized by the fluid shear and bulk viscosity. Although conceptually transparent, the actual lattice implementation of such term meets with a number of subtleties. First, it was recognized that the term S_{ab} coincides with the hydrodynamic stress tensor, only up to a normalization constant, depending on the shear and bulk viscosities, respectively. This is because the FDT appeals to a white-noise expression on a mesoscopic scale, which corresponds to a finite-memory kernel at the microscale. Second, it was discovered that the expression (71), which only acts on the momentum-flux tensor, leads to violations of FDT at small scales, of the order of a few lattices sites. This is due to the extra-dissipation carried by high-order kinetic moments, which probe the fine-grain structure of the lattice. This problem, which can be relevant to the dynamics of fluctuations in the vicinity of solid walls, was cleverly solved by introducing an extra-term in (71), coupling explicitly to high-order kinetic moments (25). This heuristic procedure has been subsequently explained by a very elegant treatment of the statistical dynamics foundations of FLB (26). The FLB theory is still in flux, expecially with concern to its application to strong non-equilibrium situations, such as nano-confined fluids under shear. This is a difficult tangle of physics and numerics, because under such conditions it is the FDT itself which goes under question on physical grounds, and requires local modifications, to the very least. On the other hand, the dynamics of fluctuations under strong confinement is exposed to a number of technical issues, such as the specific implementation of the boundary conditions, the nature of the collision operator (single versus multiple-time relaxation), to name but a few. In addition a more robust time marching procedure would also be needed, to operate the scheme under a strong-fluctuating regime. So, far, FLB is indeed limited to weak fluctuations, with noise/signal ratios of the order of $\theta \equiv kT/mc_s^2 \sim 10^{-4}$. This is partly inherent to the one-body character of the distribution function, but can probably be significantly raised by proper technical upgrades. At the moment, it appears like locally adapted multi-time collision operators, carrying a dependence on the local fluid speed and possibly stress tensor as well, might lead to improved stability. This remains an open front of LB research in nanofluids.

7.4 LB summary

LB is a sort of heterotic hybrid between particle and grid methods. Since it is based on lattice-bound particle trajectories, it inherits the efficiency of particle methods. In particular, the streaming step is an exact operation on the lattice because particles hop from site to site with zero loss of information. On the other hand, by propagating a probability rather than a black/white presence/absence information as in LGCA, LB imports the resolution power of field-grid methods. In a nutshell, it is a smooth field moving along straight trajectories. This stands in stark contrast with the fluid dynamic representation, in which any physical quantity, including the fluid momentum itself, is transported along material fluid lines, which can develop wild space-time dependencies in complex flows. Such space-time dependencies can significantly undermine the stability of the scheme. On the other hand, despite this simplicity, the LB is capable of reaching very small viscosities, because local equilibria are not realized through collisions, but prescribed at the outset. This property is unique to LB within mesoscopic fluid methods. Of course, despite its mounting success, LB is not free from pitfalls and limitations; LB formulations on non-cartesian grids are typically less efficient than the standard version, which can prove to be a stringent limitation for many engineering applications dealing with real-life geometries. Also, as previously discussed, computing flows with strong heat exchange and compressibility effects, still raises a very serious challenge to LB methods.

8 Summary

Summarizing, the last two decades have witnessed a vigorous growth of mesoparticle methods for the simulation of complex fluid flows. These methods stem from a completely new conceptual paradigm to the physics of fluids: statistical physics rather than continuum mechanics. Mesoparticle methods appear well placed to tackle problems for which the continuum fluid equations are called into question on physical grounds, such as micro and nanoflows, or prove simply too complex to be solved efficiently by discretization of continuum partial differential equations. Multiphase, multicomponent, colloidal and polymer flows at low-Reynolds number and in complex geometries are typical examples in point. Owing to its particle-field nature, the lattice Boltzmann method can handle the above complexities also in conjunction with high Reynolds flows. One of the most promising directions for future research in the field concerns the development of multiscale methods, combining the strengths of each method. This will permit to tackle a variety of complex problems at the interface between fluid dynamics, condensed matter and material science, with numerous applications in many fields of science and engineering.

9 Acknowledgements

Fruitful discussions with P. Espanol, A. Greiner D. Kauzlaric and J. Korvink are kindly acknowledged.

Bibliography

[1] Coarse Graining of Condensed Phase and Biomolecular Systems, G. Voth editor, CRC Press, 2009; P. Castiglione, M. Falcioni, A. Lesne and A. Vulpiani, Chaos and Coarse Graining in Statistical Mechanics, Cambridge Univ. Press, 2008

[2] B. Boghosian and S. Succi, in preparation

[3] M. Briscolini et al, Phys. Rev. E, 50, 1745, (1994)

[4] E. Flekkoy and P. Coveney, Phys. Rev. Lett., 83, 1775 (1999)

[5] R. Ruud and J. Broughton, Phys. Rev. B, 58, (1998)

[6] D. Kauzlaric, P. Espanol, A. Greiner and S. Succi, Macromol. Theory and Simulation, 20, 526 (2011)

[7] R. Zwanzig, Non-equilibrium Statistical Mechanics, Oxford Univ. Press, 2000

[8] J. Hoogerbrugge and J. Koelman, EPL 19, 155, (1992)

[9] I. Vattulainen et al J. Chem. Phys., 11, 3967 (2002)

[10] P. Espanol and P. Warren, EPL 30, 191, (1995)

[11] R.D. Groot and P. Warren, J. Chem. Phys. 107, 4423, (1997)

[12] E. Moeendarbary, T.Y. Ng and N. Zangeneh, Int. J. of App. Math. 1, 737, (2009)

[13] B. Chopard and D. Droz, Cambridge Univ. Press, (1999)

[14] U. Frisch, B. Hasslacher, Y. Pomeau, Phys. Rev. Lett. 56, 1505, (1986)

[15] S. Succi, The Lattice Boltzmann Equation, Oxford Univ. Press, (2001); R. Benzi, S. Succi, M. Vergassola, Phys. Rep. 222, 145, (1992); S. Chen and G. Doolen, Annu. Rev. Fluid. Mech., 30, 329 (1998); C. Aidun and Annu. Rev. Fluid. Mech., 42, 439 (2010);

[16] S. Succi, Europ. Phys. J. B, 64, 471 (2008)

[17] Y.H. Qian, D. d'Humieres and P. Lallemand, Europhys. Lett., 17, 479 (1992)

[18] I. Karlin, A. Gorban, S. Succi and V. Boffi, Phys. Rev. Lett., 81, 6 (1998)

[19] X. Shan and H. Chen, Phys. Rev. E, 47, 1815 (1993)

[20] F. Alexander, S. Chen and J. Sterling, Phys. Rev. E, 53, 2289, (1993)

[21] X. He, S. Chen and G. Doolen, J. Comp. Phys., 146, 282, (1998)

[22] M. Sbragaglia et al, J. Fluid Mech., 628, 299, (2009)

[23] A. Ladd, J. Fluid Mech., 271, 285 (1994)

[24] Tony was kind enough to send me an email, asking whether, in my opinion, adding noise to LB would be an interesting move. My answer was very skeptical, a big mistake I sorely regret. Good enough that Tony did not heed at my skepticism, and went on to develop his beautiful work...

[25] R. Adhikari et al, EPL 71, 473, (2001)

[26] B. Duenweg and A. Ladd, Adv. Polym. Sci., 221, 89, (2009)